TO THE EDGE OF
THE UNIVERSE

TO THE EDGE OF
THE UNIVERSE
THE EXPLORATION OF OUTER SPACE WITH
NASA
INTRODUCTION BY KEVIN KRISCIUNAS

Exeter Books

NEW YORK

A Bison Book

First published in U.S.A.
by Exeter Books
Distributed by Bookthrift
Exeter is a trademark of Bookthrift Marketing, Inc.
Bookthrift is a registered trademark of Bookthrift Marketing, Inc.
New York, New York

ISBN 0-671-08196-9

Printed in Hong Kong

Reprinted 1987

Photo Credits

**The National Aeronautics and Space Administration
(NASA)** provided all photographs and illustrations
except those indicated below.
Boeing Aerospace Company 98-99
Hughes Aircraft Company 56, 83
**Kitt Peak National Observatory via Hansen
Planetarium** 10-11, 14-15, 27, 28-29
© **Kevin Krisciunas** 24-25, 26-27, 30-31, 32
© **Bill Yenne Collection** 9(both), 12, 12-13, 16, 17,
18-19, 31

Designed by Bill Yenne

Edited by Bill Yenne and Susan Garratt

Captioning by David Pahl

NASA Space Science and *Pioneers In Space* articles
by Victor Siegel, NASA

We would especially like to thank Sean Martin of the
Hansen Planetarium; Steve Pehanich of the Lockheed Missiles & Space Company; Duane Brown and
William Marshall of NASA Headquarters; Jurrie
van der Woude and Mary Beth Murrill of the NASA
Jet Propulsion Laboratory; Peter Waller of the
NASA Ames Research Center and David Thomas of
the NASA Goddard Space Flight Center for their
valuable assistance in making this book possible.

Mr Krisciunas also wishes to thank Dr T R Geballe
for reading the *New Age Of Astronomy* manuscript
and for making numerous suggestions.

(*Page one*) Mariner 1 was the first in a series of
spacecraft designed for interplanetary exploration.
This artist's conception shows the craft as it flies by
Venus.

(*Pages 2 and 3*) This spectacular view of Saturn's ring
system was recorded by Voyager 2 with clear, orange
and ultraviolet frames that later were highly enhanced
by computer-processing techniques.

(*These pages*) A view of the southern hemisphere of
Venus as seen by Mariner 10 through ultraviolet filters.
The pattern of dots is on the face of the camera for
calibration purposes and can be removed by computer
processing.

Contents

The New Age of Astronomy   6

NASA Space Science   34

Pioneers in Space   64

The Hubble Space Telescope   176

Index   190

The New Age Of Astronomy

The desire to understand and explore the universe is as fundamental to humankind as breathing. One only needs to behold the star-spangled firmament on a clear, moonless night to understand why astronomy is the oldest and most honored of the sciences. The study of astronomy straightaway demonstrates how vast and varied are the celestial wonders—silently spinning icy planets, luminescent clouds of gas and dust where stars are born, pinwheel galaxies, colliding galaxies and clusters of galaxies that show that the universe on its largest scale is like a series of interlocking bubbles. Our observing platform and home, Earth, that small blue and white planet third from the Sun, is insignificant in this cosmic immenseness. Ours is a very young civilization, one measured in thousands of years, while the age of the universe is measured in billions of years. It is with a sense of wonder and humility that we have contemplated the stars for centuries. In the past 30 years we have expanded the scale of our quest for an understanding of the universe by actually visiting other worlds.

As Columbus, standing on the shores of Spain, may have sensed a New World beckoning, so we stand at a new frontier that beckons us. When viewed at any time in history by a contemporary, the recent efforts to understand the universe have always loomed large in relative importance. However, it can safely be said that our present efforts are child's play compared to the accom-

plishments we know can be achieved in the near and distant future, since new equipment and techniques are continuously being developed. Ours is the latest new age of astronomy, but it is an astronomy continuously being regenerated.

In the past, astronomical progress has not always been so energetic, yet viewed in the context of their times, the discoveries and accomplishments of ancient and historical astronomers were indeed remarkable. Our knowledge of the astronomy of the ancients is scanty, but we do know that the Chinese made records of eclipses as early as 2300 BC. Chinese records of supernovas and comets have particular significance for modern astronomy. Rudiments of astronomical knowledge were used by the Egyptians for astrological purposes and for aligning their pyramids. The

(*Above*) **The Calendar Stone of the Aztecs provided a means of matching the seasons with the movements of the stars and planets. At the center is the sun god surrounded by the symbols of earth, air, water and fire. Encircling these are twenty precise divisions representing days of the month. The Aztecs recognized a 365 day year divided into eighteen months of twenty days each. The remaining five days of the year were viewed as unlucky by the superstitious Aztecs.**

Babylonians gave us many of the constellations that we still use, and it is safe to say that the most advanced ancient astronomy, that of the Greeks, was directly derived from Babylonian models.

The two most significant ancient astronomers were Hipparchus (2nd century BC) and Ptolemy (2nd century AD). Hipparchus was an observational astronomer who lived on the island of Rhodes. He compiled a catalog of 850 stars and made many accurate mea-

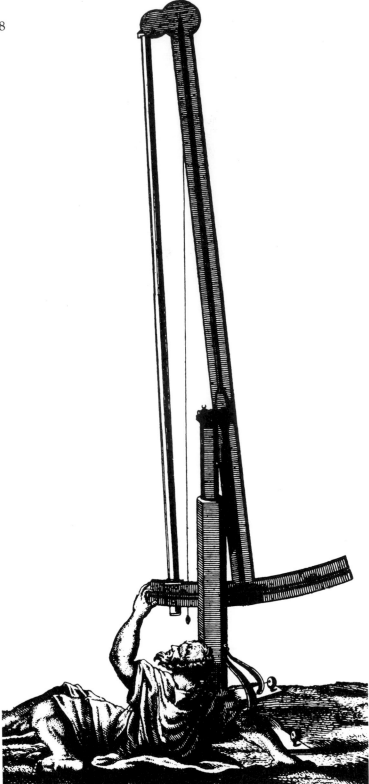

Galieo's thirty-power telescope was a simple affair consisting of a large convex lens at the front and a smaller concave lens for an eyepiece. Although this telescope was inefficient by our standards, this great advance by Galileo revolutionized astronomy and opened the heavens to the early astronomers. For the first time scientists could see more than the brightest stars.

surements that allowed the positions of the Sun, Moon and planets to be calculated in advance. Ptolemy, who lived in Alexandria, was a theoretician by the standards of today. He produced the most elaborate Greek mathematical model of the motions of the heavenly bodies, based primarily on the observations of Hipparchus and others. Though astronomers such as Aristarchus (3rd century BC) had suggested —correctly—that the earth and planets all orbited the Sun, Ptolemy's work, which summarized centuries of Greek developments, was based on the notion that the earth was the center of the universe. We criticize Ptolemy today for having presented the solar system inside out, but from the standpoint of his era he did the best he could. Definitive evidence

for a moving Earth was only obtained in the eighteenth and nineteenth centuries.

The Dark Ages were not dark! This goes for astronomy as well. While there was little Greek and Roman astronomy being pursued, it was the golden age of Arabic astronomy. At places like Baghdad, Cairo, Maragha and Samarkand there were notable schools of astronomy from the ninth through the fifteenth centuries. Whereas the Greeks had observed with instruments made of wood or stone, the Arabs made substantial advances in metal working, giving rise to more accurate observational tools. They also built on Ptolemy's mathematical model. Some even provided evidence in favor of a Sun-centered solar system.

The Polish cleric Nicolaus Copernicus is regarded as the founder of modern astronomy. His book *On the Revolutions of the Heavenly Spheres,* published in 1543, showed for the first time that the mathematical model of the motions of the planets could be simplified by displacing the earth from its previously allocated position in the center of the universe. 'In the center of all rests the Sun,' wrote Copernicus. 'For who would place this lamp of a most beautiful temple in a better place?'

Copernicus, like Ptolemy, was a theoretician and calculator, not an observer. Evidence for Copernicus' theory was right around the corner. In 1608 or 1609 the telescope was invented, which, in the hands of the great Italian scientist Galileo Galilei, revolutionized astronomy overnight. Galileo's equally important contemporary, Johannes Kepler, described the telescope as a 'much-knowing perspicil, more powerful than any scepter! He who holds thee in his right hand is a true king, a world ruler…'. Here at last was a new window on the universe, a device that could make objects appear closer and allow us to see stars fainter than can be seen by the naked eye. In short order Galileo established that the Moon had mountains, the Sun had spots, Venus exhibited phases like the Moon (evidence that it orbited the Sun) and that four tiny moons orbited Jupiter (proving that not all celestial bodies have to orbit the earth). Unfortunately, Galileo consequently made many enemies in the religious ranks and was admonished to cease advocating the truth of the Copernican doctrine. He reminded others that 'the Bible teaches us how to go to heaven, not how the heavens go.'

Galileo's telescopes were very simple, each consisting of a large convex lens up front and a smaller concave lens as an eyepiece. His most powerful telescope magnified 30 times. A Galilean refractor gives colored fringes around each image, which are pretty at first sight, but which are symptomatic of a waste of light because light of each color comes to a focus at a different position. This is called chromatic aberration. A partial solution to the problem was obtained by the Englishman John Dollond, who in 1758 produced the first two-element objectives, which bring the red light and blue light to the same focus while leaving only the yellow light a little out of focus.

Another problem of the early refractors was spherical aberration. It was known how to make lenses that had spherical surfaces. In addition to the color problem mentioned the light would still not come to a perfect focus unless the telescope objective had a parabolic surface. Opticians found that if the focal length of the telescope was very long, the effect of the spherical aberration was less pronounced. By the late seventeenth century astronomers in Europe were using telescopes up to 150 feet in length supported by huge towers or masts. This gave better images and very high magnification, allowing better close-ups of the planets, but such telescopes were extremely unwieldy.

The invention of the reflecting telescope by Isaac Newton in 1668 solved the telescopic problem of chromatic aberration. While a prism or lens breaks up white light into colors, reflection off a mirror does not. Newton's reflecting telescope consisted of a concave mirror at the bottom of the telescope tube, a flat mirror farther up the tube near the mirror's focus that sent the light out the side of the tube to the eyepiece. This type of telescope, called a Newtonian reflector, is still very popular today among amateur astronomers.

For nearly 200 years the reflectors used mirrors made of special metallic alloys, similar to the material used in making church bells. They could be cast to the desired size and ground and polished to give good images. The champion telescope maker in this regard was William Herschel, the discoverer of the planet Uranus in 1781, who made hundreds of telescope mirrors in his lifetime. His largest reflecting telescope had a 48-inch-diameter mirror and a tube 40 feet long. While most astronomers were using specially made refractors with lenses a few inches in diameter to make accurate measurements of star positions, Herschel's telescopes allowed him to survey the sky to a greater depth than anyone else. The light-gathering power of a telescope is measured by the area of the main objective, be it a lens or mirror. Because Herschel's telescope mirrors were so much larger than the largest lenses that could be made at the time, he could see much fainter, hence much more distant, objects. He was the first to glimpse four of the smaller moons of Saturn and Uranus and he discovered thousands of double stars and nebulas. In effect, he founded galactic astronomy.

The problem with metal mirrors was that they tarnished rapidly. The mirrors had to be frequently refigured and repolished, a trial and error procedure at best. The first silver-on-glass telescope mirrors were made in 1856. Although these tarnished too, their shine lasted longer. Eventually, aluminum coatings were developed, but even modern telescopes need to be realuminized every one or two years.

The work horses of nineteenth-century astronomy were long-focus refractors, the largest being the 40-inch refractor of Yerkes Observatory in Williams Bay, Wisconsin. The nineteenth century was an age when astronomers still used their own eyes to look through telescopes. They measured positions of individual and

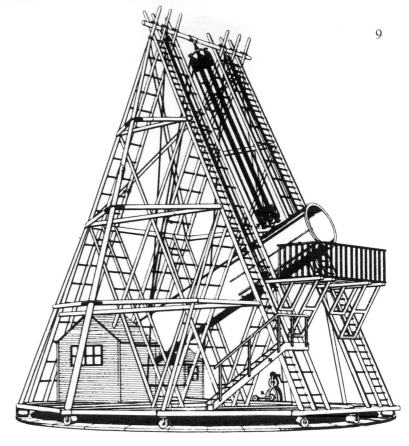

Herschel's 40-ft telescope (*above*) was inscribed to his patron King George III and required a 48-in diameter mirror and a heavy scaffold with block and tackle to raise and lower the system in order to track the stars and planets. (*Overleaf*) A view of the heavens which could only have been imagined by Herschel: The Lagoon Nebula in Sagittarius as seen from a photograph through the 3.8-meter telescope at the Kitt Peak National Observatory in Arizona.

A visual spectroscope (*below left*) and a polarizing eyepiece (*below right*) are examples of typical nineteenth century astronomical equipment. The development of astronomical spectroscopy, combined with advances in photography, led to a revolution in the science of astronomy perhaps more profound than the introduction of the first telescope. When light is passed through a prism it breaks down into its constituent colors. In the mid-1800s scientists determined that specific substances give rise to specific color lines spectra. Spectroscopy uses this tendency of light by observing the spectra of the stars and planets to determine their compositions.

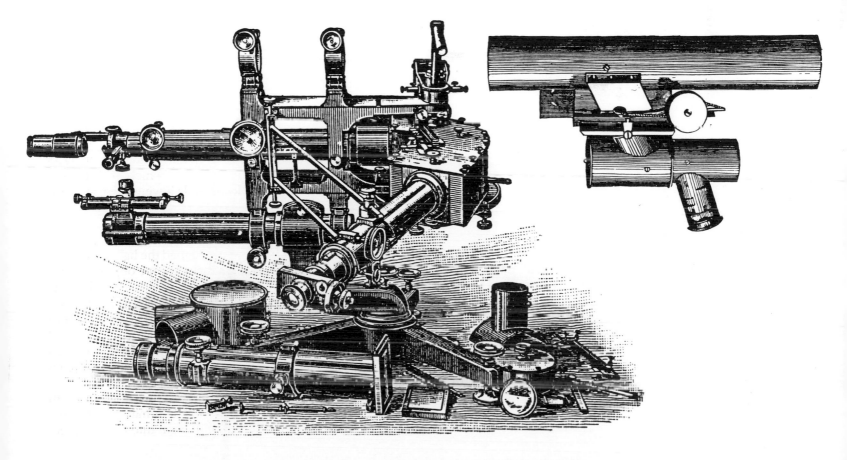

The Equatorial telescope at Washington (*left*). Progress in the science of astronomy in the early part of the twentieth century depended upon the development of larger telescopes and observatories such as the Griffith Observatory (*right*) at Los Angeles. (*Overleaf*) Advances in astronomical equipment provide us with spectacular views of the heavens. This Kitt Peak National Observatory photograph through its 3.8-meter telescope is of the Great Nebula in Orion.

multiple stars, made drawings of planets and nebulas, and derived means by which to measure the scale of the solar system and the distances to the nearby stars. By the end of the nineteenth century there were photographic emulsions good enough to allow astronomers to photograph thousands of stars in individual star clusters. This led to plans for photographically mapping the whole sky, in order to give permanent, unbiased records of the state of the celestial wonders. Photography also allowed, often by serendipity, the discovery of numerous asteroids and variable stars (stars whose light output is not constant). Most significantly, the development of astronomical spectroscopy, combined with photography, led to another revolution in astronomy as profound as the first use of the telescope by Galileo. A mid-nineteenth century physicist had once exclaimed, 'What the stars are, we do not know and will never know!' But the spectroscope and its resulting new branch of science, astrophysics, showed how wrong this opinion was.

The same 'flaw' of lenses that gives rise to chromatic aberration —the colored fringes around stars—gives us the benefit of spectroscopy. White light is broken up by a prism into its constituent colors, as Newton discovered. In the early nineteenth century it was discovered that by placing a slit in front of the prism one could view the Sun's spectrum as a series of rainbow colors, but with many dark lines superimposed on those colors. The German optician Joseph von Fraunhofer catalogued these lines, but it was not until 1859 that fellow Germans Gustav Kirchhoff and Robert Bunsen discovered the true nature of the lines. They were caused by specific substances in the Sun's photosphere, each of which gives rise to a specific set of lines. One only had to carry out some laboratory experiments to see which lines are associated with which substance, then one could observe the spectra of stars and determine their various compositions.

As is always the case, science is more complicated on closer inspection. By the 1920s it was shown how different stellar spectra result from different conditions of temperature and pressure, not just composition. It was discovered that stars are mostly made of hydrogen and helium. By the first third of the twentieth century astronomers had discovered that the Sun lay in the plane of a spiral galaxy, our Milky Way, but nowhere near its center as had been previously thought. Instead, we are some 30,000 light-years from the center of the galaxy, and our galaxy rotates once every 250 million years. By the late 1930s physicists discovered that the cores of stars are nuclear furnaces, with temperatures measured in the millions of degrees—so hot that smaller atoms are fused together into heavier and heavier atoms. The more massive stars of times gone by have long since blown off their outer shells, repopulating the intersteller medium with those heavier atoms, such that the initial composition of new stars changes over time. It is an absolute fact that the majority of the atoms in your body were once on the inside of a massive star that blew up.

The progress of astronomy in the twentieth century hinged on the creation of larger telescopes. George Ellery Hale, after founding the Yerkes Observatory, went to Southern California, where he built the Mount Wilson Observatory. A 60-foot solar telescope tower was put together, and also a 150-foot one. For nighttime astronomy a 60-inch reflector and the great 100-inch Hooker reflector were constructed. Hale's efforts at astronomical empire building culminated in the 200-inch reflector at Palomar Mountain, completed in 1948, 10 years after his death.

It was with the 100-inch reflector that Edwin Hubble discovered the general expansion of the universe. Using spectra of distant galaxies to measure their velocities along the line of sight, Hubble discovered that almost all galaxies were receding from us, and at a rate proportional to distance. From the rate of expansion it is possible to calculate the age of the universe, which modern estimates place in the range of 10 to 20 billion years.

Though infrared and ultraviolet rays had been known about since the early 1800s, and radio waves and X-rays had been discovered by the end of the century, astronomers had confined their studies to the optical window of the universe—the narrow band of visible light from the blue to the red. The invisible wavelengths of the electromagnetic spectrum include, from shortest to longest, gamma rays, X-rays, ultraviolet, infrared and radio waves. Visible light falls between ultraviolet and infrared. The first nonoptical branch of observational astronomy to be developed after World War II was radio astronomy, which can be carried out at sea level and is best conducted in the absence of electrical disturbances, far from sources of man-made radio interference. Karl Jansky, a Bell Telephone engineer, had built the first radio telescope and in 1932 discovered radio waves from the Milky Way. But he had only one scientific 'son,' Grote Reber, who built a 31½-foot-diameter dish in his back yard in Wheaton, Illinois. Reber confirmed Jansky's findings and also discovered a number of discrete radio sources in the sky.

After Jansky's and Reber's initial successes in the United States, the centers of radio astronomy were Australia, Great Britain, and The Netherlands. Radio observations led to several discoveries including quasars, whose nature is still not understood; pulsars, which are rapidly rotating neutron stars, remnants of stellar explosions; and certain galaxies that can be viewed at enormous distances because of their unusually great brightness at radio

wavelengths. Radio astronomers have mapped out the distribution of hydrogen and other gases throughout our galaxy. They have discovered that the interstellar medium is a special chemistry laboratory where molecules grow. Perhaps the building blocks of life arose in the chemical reactions taking place on the surfaces of interstellar dust grains.

The study of ultraviolet, X-ray, gamma-ray and some infrared sources has required rockets and artificial satellites. For the development of rockets and satellites let us turn back the clock 30 years. Both the United States and the Soviet Union had announced plans to send artificial satellites into orbit during the International Geophysical Year of 1957–58. The Soviet Union accomplished this first when it launched Sputnik in 1957, and that led to an unprecedented commitment on the part of the US government to fund science and technology. No longer were we confined to platform Earth. The door to space was wide open.

The Soviet-American space race, while motivated in large part by political concerns, led to a new phase of solar-system astronomy that has largely separated itself from the rest of astronomy; it involves not so much telescopes as cameras, particle analyzers, and human-life support systems. This has given us breathtaking close-up views of the Moon and planets, and has allowed us to land men on the Moon and send robot landers to Venus and Mars. Robot emissaries from our planet are flying into interstellar space carrying information about their human creators. Since many of these developments unfolded before the constructon of full-fledged orbiting astronomical observatories, we shall turn to them next.

The solar system, including the earth and its moon, are about 4.6 billion years old. Most of the Moon's present surface was formed in the first 600 million years of its history. But because the moon possesses no atmosphere and has no flowing water or other

liquids on its surface, there is essentially no erosion. With his simple refractor Galileo could tell that the Moon was not a smooth sphere, but rather was badly scarred, cratered and pockmarked. One can easily imagine the deluge of boulders and small asteroids that will impact the Moon in the near and distant future. (The Earth has been just as heavily bombarded, but because of the extent of our oceans and the erosion of land areas, the evidence of meteoric impacts is harder to find.)

For the Moon it was just another small impact on 13 September 1959, but to earthlings it was a milestone. The Soviet probe Luna 2 became the first man-made object to have physical contact with another celestial body when it landed some 500 miles north of the visual center of the moon, near the crater Archimedes. A number of American and Soviet probes followed. The Soviet Luna 3 obtained the first photographs of the back side of the Moon in October 1959. In February 1966 Luna 9 became the first probe to soft land on the moon. The United States had its share of successes and failures as well. Rangers 3, 4 and 5, launched in 1962, sent back some close-up pictures of the Moon. In 1964 and 1965 Rangers 7, 8 and 9 each sent back thousands of extremely high-resolution images before their respective impacts. All of this information was crucial to finding a safe landing spot for the future manned visits. How well I remember watching Ranger mission control headquarters as the latest satellite beamed back ever sharper images of the fast-approaching lunar surface. The multi-million dollar probe then smashed into the Moon's face, sending a thousand pieces flying. There was no sound on the Moon, but the engineers at mission control, sporting crew cuts, white shirts, and skinny dark ties, cheered in exhilaration as their satellite annihilated itself on that very large rock in the sky.

The American probe Surveyor 1 soft landed on the Moon on 2 June 1966, four months after Luna 9 and after Luna 10 became the first successful lunar orbiter. Surveyors 3, 5, 6 and 7 soft landed in 1967 and 1968. Other successful Soviet landers and orbiters were joined by American orbiters, then by the first manned lunar orbital flight of Apollo 8 in late December of 1968. When Frank Borman, James Lovell, and William Anders went into their first orbit around the Moon, it marked the first time that any human had been cut off from all contact with the earth. As they emerged from the far side of the Moon, a crescent Earth rose in front of them, and millions of viewers followed their progress on television and radio that Christmas Eve morning long before dawn.

Apollo 9 astronaut Rusty Schweikert had this to say about viewing the earth from space:

It is so small and so fragile and such a precious little spot in that universe that you can block it out with your thumb, and you realize that on that small spot, that little blue and white thing, is everything that means anything to you—all of history and music and poetry and art and death and birth and love, tears, joy, games, all of it on that little spot out there that you can cover with your thumb. And you realize from that perspective that you've changed, that there's something new there, that the relationship is no longer what it was....

And when you come back there's a difference in that world now. There's a difference in that relationship between you and that planet and you and all those other forms of life on that planet, because you've had that kind of experience....

And all through this I've used the word 'you' because it's not me—it's you, it's we. It's life that's had that experience.

(Right) Two NASA spacecraft on the surface of the moon in November of 1969. In the foreground is Surveyor 3 which landed on the lunar surface in April of 1967 and paved the way for man to explore the surface of the moon. In the background, just 600 feet from Surveyor 3 is the Apollo 12 Lunar Module which had just landed with Alan Bean and Charles Conrad aboard.

Apollo 11 was the first manned landing on a celestial body other than the earth. Everyone knows Neil Armstrong's first words as he stepped onto the lunar surface on 20 July 1969. But what was the first word broadcast from the Moon before the descent of the ladder? It was 'Houston.' Buzz Aldrin had radioed: 'Houston, Tranquility Base here. The Eagle has landed.'

There were six successful manned missions to the Moon, the last being Apollo 17 in December 1972, which carried a geologist, Harrison Schmidt, on board. Apollo missions had proven that humans could leave that cradle, Earth, venture to new worlds, and even bring back hundreds of pounds of lunar rocks for study. This was no longer astronomy, but a grand field trip, holding all the potential for a colonizing scheme for space.

The American Pioneer 5 was the first interplanetary probe. Launched in March 1960, it transmitted to a distance of 23 million miles. Pioneers 6 through 9, launched between 1965 and 1968, studied the Sun. Pioneer 10 (March 1972 through December 1973) and Pioneer 11 (April 1973 through December 1974) made the first two flybys of Jupiter, revealing in astounding detail that giant planet's Great Red Spot like an eye of a cosmic whale staring back at us.

Mariner 2 flew by Venus in December 1962. Mariner 4 made the first successful flyby of Mars in July 1965, showing craters similar to those of the Moon and measuring the thin Martian atmosphere. Other Mariners spied on Venus and Mars, including Mariner 9, the first man-made Martian orbiter, which returned thousands of pictures of the surface and its tremendous sand storms. Mariner probes gave us the first close-ups of the two Martian moons, Phobos and Deimos. Mariner 10 made the first of three swings by Mercury in March 1974.

As was the case with the lunar probes, flybys, orbiters and crashes were followed by soft landings on other planets, in situ surveys of the landscapes and samplings of the planetary soils. The Soviets achieved the first soft landing on another planet in December 1970 when Venera 7 survived the descent through the hellish atmosphere of Venus to a surface even more hellish. Many Soviet Venus landing probes followed, along with the Americans' Pioneer-Venus mission, each revealing a planet whose mostly carbon dioxide atmosphere has a pressure 90 times that of Earth's and where the surface temperature is hot enough to melt lead.

While Venus has received mostly Soviet-built vehicles, the outer planets have been the province of the American craft. Vikings 1 and 2 landed on Mars in the summer of 1976, showing a Martian landscape not unlike the sunset-red terrain one sees from the summit of Mauna Kea in Hawaii. Dashing the hopes of many, Viking failed to find evidence of life on Mars. However, other discoveries about Mars have led to speculation about past life there.

The two Voyager probes have widened our horizons more than any. Launched in the late summer of 1977, both visited Jupiter and Saturn in 1979. A tenuous ring was discovered encircling Jupiter. The four Galilean moons, dwarfed by their parent planet, are as large and as varied as Pluto and Mercury, and are shown to be equally bizarre. Europa is like a marble egg covered in cracked ice. Io has the most active volcanoes of any object in the solar system. The Saturn flybys revealed thousands of ringlets making up that planet's famous ring system. Enigmatic bright or dark spokes were seen in the rings. The F ring was seen to be braided, and new Saturnian moons were found. Voyager 2 is completing its grand tour of the solar system, having passed Uranus in January 1986

(*Right*) **Apollo 17 Geologist Harrison Schmitt working near the Lunar Rover in the Taurus Littrow Region of the moon in December of 1972. This area proved to be of great geological interest for scientists because of the surface coloration. As can be seen, the surface of the moon is uniformly grey with the exception of the reddish dirt, probably an iron oxide, found on the small knoll to the right of the Rover.**

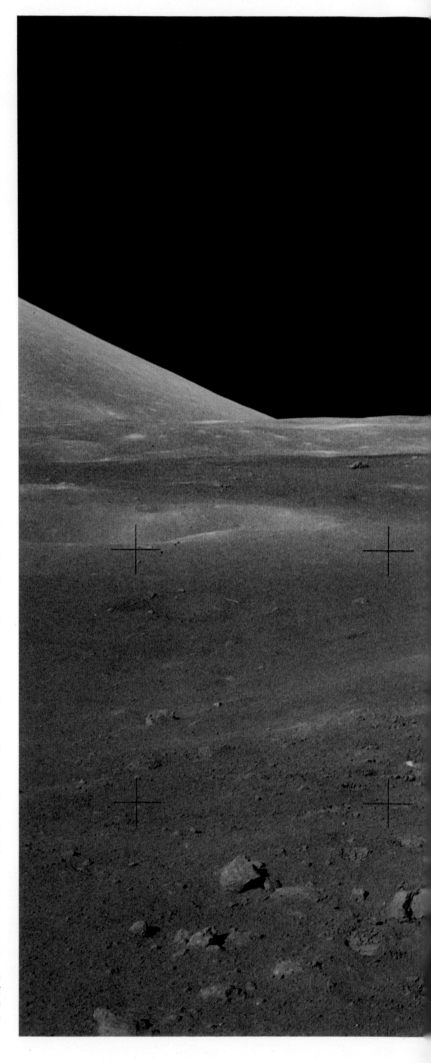

and heading on to Neptune for a rendezvous in 1989. Uranus and Neptune can be imaged by satellite far better than by the telescopic behemoths now being panned for Earth, or even by the Space Telescope, soon to be launched by the Space Shuttle astronauts.

The Space Telescope will be the most advanced and most expensive ($1 billion) of the dozen or so orbiting astronomical observatories launched since 1966. Before we discuss its great scientific promise, let us briefly review its more humble predecessors. They can be divided into two basic classes—analytical instruments, which are used to make observations of already known objects; and survey instruments, which are systematically used to map out the sky at previously unknown thresholds of brightness or resolution. Some of the orbiting observatories can be used in more than one mode; once new sources are discovered other detectors can be used to provide more detailed information on especially interesting targets.

What differentiates the manned and unmanned space missions from the orbiting observatories is that the former were used to study one body of the solar system at a time, taking pictures, measuring gases and magnetic fields, and gathering substances from the actual bodies. The orbiting observatories cannot reach the actual stars or galaxies under study, so must be more passive in their approach. Like ground-based observatories, they must be content with whatever light waves the astronomical objects have sent their way when those observations are made.

Because orbiting observatories are above Earth's atmosphere, they can receive all wavelengths of light. It is always clear up there, so there are no cloudy nights to disappoint the observer. These advantages outweigh some of the constraints of cost and accessibility, of which there are naturally a few. Satellites generally cost hundreds of millions of dollars, rather than 'only' millions. If something breaks on the satellite, it might not be fixable by remote control. Furthermore, if the satellite closely orbits the earth, the earth can get in the way. One has to make sure that the telescope is not pointed at the Sun (unless, of course, it is a solar telescope). Furthermore, the earth's atmosphere filters out radiation such as ultraviolet and X-rays, which are harmful to living organisms, and the earth's magnetic field as well as upper atmosphere may affect observations. The bottom line is that one must go into orbit to investigate the universe at gamma-ray, X-ray, most ultraviolet, and many infrared wavelengths.

There were very few infrared astronomers 20 years ago, but now there are many. For their work the primary constraint is water vapor in the earth's atmosphere. The higher a telescope is above sea level, the less water vapor there is to look through. To observe celestial bodies at many wavelengths this required the invention of new kinds of detectors often placed on high mountaintops or even launched into space by rockets. Mauna Kea on the island of Hawaii, at an altitude of 13,800 feet above sea level, is one of the most significant places where infrared astronomy is being studied. It is twice as high as any other major ground-based observatory, allowing observations at infrared and submillimeter wavelengths that can be carried out nowhere else. At present the world's seventh, eighth and twelfth largest telescopes are to be found at the summit of Mauna Kea. (They are the United Kingdom Infrared Telescope, the Canada-France-Hawaii Telescope and the NASA Infrared Telescope Facility, respectively.)

(Left) **The observatories of Mauna Kea on the island of Hawaii. A four-nation cooperative cluster of facilities, we see the NASA Infrared Telescope at right, the Canada-France-Hawaii Telescope at center and the United Kingdom Infrared Telescope on the left. These infrared telescopes are able to 'see' objects which are invisible to optical telescopes because infrared radiation passes freely through the interstellar dust clouds which block visible light.**

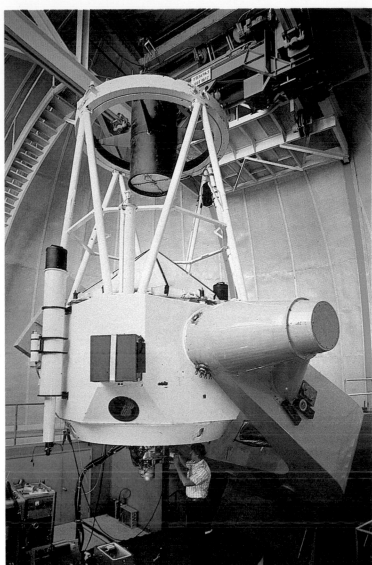

The Kitt Peak National Observatory (*left*) is located at an elevation of 6900 feet at Kitt Peak, Arizona. The observatory was completed in 1973, and its astronomical equipment includes both a 2.1-meter (*above*) and a 3.8-meter telescope. The resolution of these powerful telescopes may be seen in the view (*overleaf*) of the Eagle Nebula in the Serpens Constellation in the Northern Hemisphere. The 3.8-meter Nicholas U. Mayhall Telescope at Kitt Peak National Observatory is the seventh largest ground-based telescope in the world.

Higher still than Mauna Kea, but no longer ground based, is NASA's Kuiper Airborne Observatory (KAO), a modified C-141 four-engine military transport jet that carries a 36-inch telescope to altitudes as high as 45,000 feet. Flying on the KAO is like being a low-altitude astronaut. The crew wear NASA-blue flight suits, headsets and oxygen masks. Once airborne, if something breaks, there is no repairman to call, so the crew must be able to solve a wide variety of potential problems. Major discoveries associated with the KAO include mapping of the center of our galaxy, the discovery of the Uranian ring system and the detailed study of regions of star formation.

Lower-energy photons of infrared light have also been gathered by orbiting satellites. From January to November 1983 the Infrared Astronomical Satellite (IRAS) surveyed the sky at wavelengths of 12, 25, 60 and 100 microns. Operated by the United States, United Kingdom and The Netherlands, IRAS was a 22.4-inch telescope cooled to a temperature of 2½ degrees above absolute zero. (This was done so that the telescope, which also radiates infrared waves, does not see itself.) While observations at two of the IRAS wavelengths can be made from the ground and the other two can be accessed from the Kuiper Airborne Observatory, the value of

the IRAS survey was that it gave *unbiased* maps of nearly the entire sky at the four wavelengths and at a much fainter limiting brightness than ground-based surveys.

IRAS discovered a total of five comets, 'infrared cirrus' clouds and a cloud of material around the star Vega similar to a system of planets or asteroids in the making. Star formation regions are the bread and butter of infrared astronomers, and, with the aid of computer processing, IRAS produced wonderful pictures of the Orion Nebula and other concentrations of nebulous material along the plane of the Milky Way. Other galaxies—spirals, ellipticals and irregulars—were logged as well. Having unbiased statistics on the luminosities of these building blocks of the universe tells us much about its structure and evolution. Of some 9000 IRAS sources identified in preliminary analysis, a few have no known counterparts at other wavelengths. Like some of the gamma-ray and X-ray sources, these anomalies are the constituents of minirevolutions in each field.

Ultraviolet astronomy is the astronomy of hot stars and hot gas ($10,000°$ K to $1,000,000°$ K). The Sun's chromosphere is one example of such gas. Ultraviolet spectra show spectral lines resulting from highly ionized atoms (atoms stripped of many electrons), and give us information about the processes of energy flow (hot winds) in the atmospheres of hot or active stars. Some not so hot stars are found to have large areas covered by star spots, like the spots on the Sun but more extensive. They, too, are strong ultraviolet sources. Wherever there are explosive phenomena in stars or galaxies, there are inevitably copious amounts of high-energy radiation. Black hole candidates, pulsars, exploding stars, and active galactic nuclei have been studied in the ultraviolet, X-ray and gamma-ray regions of the electromagnetic spectrum. A variety of ultraviolet telescopes, all analytical instruments, have been built in recent years.

The first Orbiting Astronomical Observatory (OAO) was launched in 1966, but it failed after two days. Its succesor, OAO-2, went aloft in December 1968. It had eleven telescopes: four were 12½-inches and four were 8-inches, and all were used to measure the ultraviolet brightnesses of hot stars over the wavelength range 1050 to 3200 Ångstroms; a 16-inch reflector for photometry of nebulas from 2100 to 3400 Å; and two small telscopes that took spectra in the 1000 to 4000 Å range. (Visual light ranges in wavelength from 3500 to 7000 Å.) A sister satellite, OAO-3, renamed *Copernicus* in honor of the 500th birthday of Nicolaus Copernicus, was launched in 1972. It operated for nine years, and carried a 32-inch ultraviolet telescope that was used to take spectra plus three small X-ray telescopes. In 1978 the International Ultraviolet Explorer (IUE) project, a joint American, British and European Space Agency (originally the European Space Research Organization) program, launched a 17.7-inch reflector designed to take spectra from 1150 to 3200 Å. It is still in operation today. Each succeeding orbiting ultraviolet telescope has exhibited improvements in limiting sensitivity. Improvements in detector technology and the telescopes themselves allow fainter stars to be measured.

More recently, in 1983, the Soviets launched Astron, a 32-inch ultraviolet telescope operating in the 1100 to 3600 Å range that also carries X-ray spectrometers. One unusual aspect of Astron is that its orbit takes it halfway to the Moon, so that it can carry out observations without interruption for many hours and without interference from the earth's magnetic field.

The first X-ray telescopes were surveyors of the skies. Not enough could be extrapolated from visual wavelengths to make good guesses as to where to look for X-ray sources other than the

(Right) Shocked modular hydrogen in the source DR 21, observed at 2.12 microns wavelength in the near infrared. The source's two lobes are shown here in red.

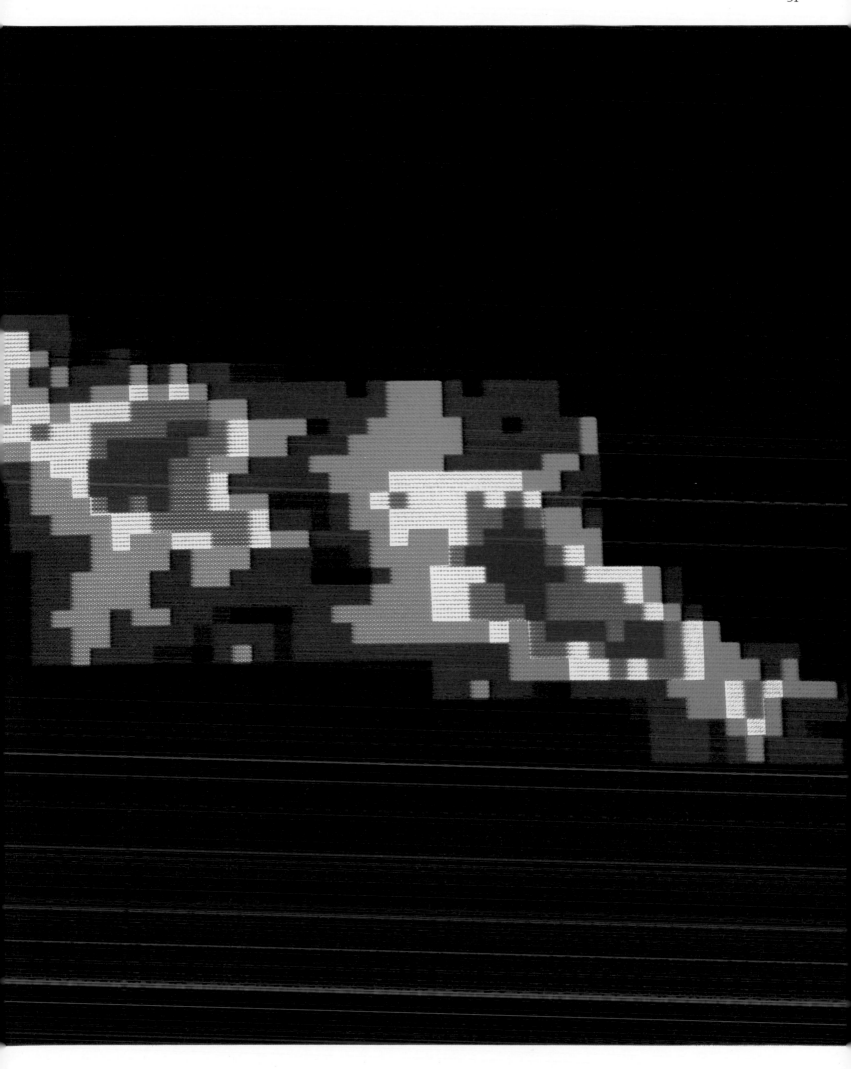

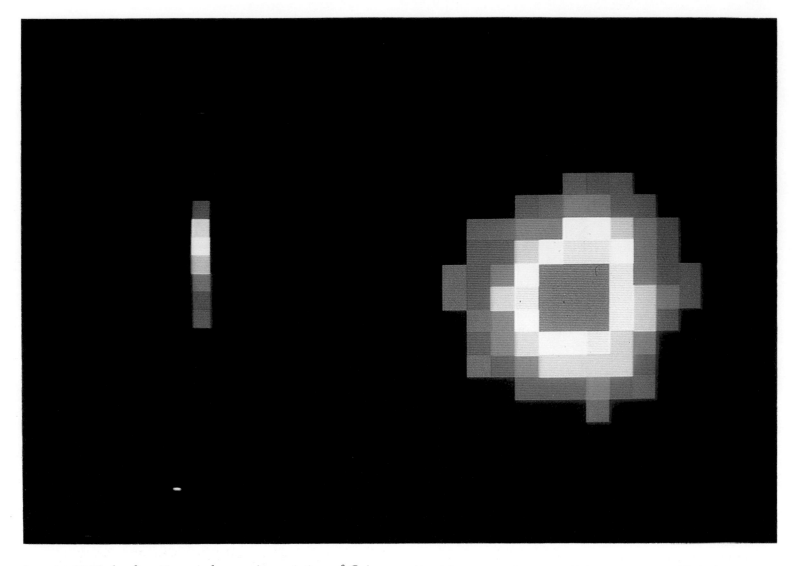

Sun. In 1962 the first X-ray 'telescope,' consisting of Geiger counters, was sent aloft by rocket. The scientists expected to find X-rays coming from the Moon. Instead they found a bright source in the constellation Scorpius, known since as Sco X-1. Subsequent rocket flights brought the total to about 30 sources.

The first orbiting observatory dedicated to X-ray astronomy was the American Small Astronomical Satellite number 1 (SAS-1), better known as Uhuru, meaning 'freedom' in Swahili. The satellite was launched in 1970 from Kenya, and its name honored Kenya's Independence Day. During its 2½-year lifetime, Uhuru mapped the X-ray sky and detected 339 X-ray sources. Most of them had never been distinguished before at any wavelength. The majority of the sources was distributed along the plane of the Milky Way and concentrated toward its center. Much more sensitive—by a factor of 1000—were the High Energy Astronomy Observatories (HEAO). HEAO-1, launched in 1977, and HEAO-2, launched in November 1978, greatly increased the number of known X-ray sources. The latter, renamed the Einstein Observatory in 1979 in honor of Albert Einstein's 100th birthday, was the first X-ray telescope to give direct images of X-ray sources. Both wide-angle shots (with a resolution of 1 arc minute, 1/60 of a degree) or close-up images with a resolution as good as that obtainable with ground-based optical telescopes could be taken. This allowed the X-ray sources to be identified with optical or radio sources (if they had optical or radio counterparts), allowing a multipronged attack to elaborate on their nature. Einstein operated until April of 1981: its noble successor was the European Space Agency's Exosat observatory, launched in May 1983. Like the Soviet's Astron ultraviolet and X-ray observatory, Exosat has an unusual orbit that takes it halfway to the Moon. Observations can be made continuously for

(*Above*) **A computer color-enhanced infrared photograph of Halley's Comet as it traveled toward Earth in December of 1985. The bright spot near the center of the body is the nucleus of the comet, surrounded by a blanket of ice. This comet travels on an elliptical orbit taking 76 years for each complete circuit.**

72 hours, and data can be accessed in as few as 10 microseconds. The Japanese also launched an X-ray satellite in 1983 known as Tenma (Pegasus). It joined their other X-ray observatory, Hakucho (Cygnus), aloft since 1979.

The findings of these X-ray satellites have been many. It was found that many red dwarf stars, normally thought to be just faint normal stars, are often prodigious X-ray emitters. Their chromospheres, heated to a million degrees, give off greater amounts of X-rays than our much brighter Sun. The source star, Cen X-3, sends out pulses of X-rays every 4.8 seconds, and the pulses turn on and off every 2.1 days. This has led to an interpretation of Cyg X-3 as a binary star containing a normal stellar companion with a 2.1 day orbital period. Another interesting source is Cyg X-1, whose X-ray pulses were discovered by Uhuru and which was first imaged by Einstein in 1978. It consists of a very hot ninth-magnitude star 25 times as massive as the Sun and a collapsed object of at least nine solar masses. Because the latter is much greater than the allowed mass of a neutron star, it is possible that it is a black hole.

Of the various flavors of electromagnetic radiation, or light waves, which only differ by their wavelengths (or hence by their energies), the most energetic (those of the shortest wavelength) are gamma-rays. When cosmic rays (which are high-energy particles) collide with interstellar gas, high-energy radiation is given off. The technology for detecting these rays is different from the method for detecting X-rays. An X-ray telescope has what looks like a nested set of metal garbage cans. The X-rays graze off the accurately figured metal surfaces and are imaged by the X-ray detectors. Gamma-rays have wavelengths much smaller than the distances between atoms in solids, and thus would be absorbed by such mirrors. A gamma-ray detector consists of a stack of flat spark chambers alternating with thin metal plates. High voltage is applied to the plates and incoming gamma-rays *or* cosmic rays cause sparks. Detecting the gamma-ray sources requires filtering out the cosmic ray flux, which outnumbers the gamma-ray radiation by a factor of 100,000 to one. In fact, in all of gamma-ray astronomy less than 250,000 photons have been measured. They come singly, not as a torrential flow like lower-energy photons.

Only two gamma-ray observatories have been used so far: SAS-2, which operated for seven months in 1972–73; and ESA's COS-B satellite, which operated from 1975 to 1982. Such telescopes could only locate sources to 2 degrees in the sky (four times the diameter of the Moon), nowhere near enough in resolution to allow certain identifications of all the sources. Only two types of gamma-ray sources have been identified for certain—pulsars and molecular clouds. The Vela and Crab pulsars also emit X-rays and radio waves. They are bona fide gamma-ray sources, too. Molecular clouds are birthplaces of stars and are the most massive known entities in the galaxy, up to one million solar masses. One puzzling gamma-ray source is called Geminga, named so because it is a gamma-ray source in the constellation Gemini. Though there is an X-ray source nearby in the sky, it does not appear to be the cause of the gamma-ray flux. Likewise, there is no radio or optical source at the most likely position. What is it? A new type of cosmic object? No one knows for sure.

To quote Robert Browning, 'A man's reach should exceed his grasp, or what's a heaven for?' Astronomers, insatiable for more powerful telescopes and desirous of opening up ever wider the aforementioned windows on the universe, have plans for major new instruments. The Hubble Space Telescope will be launched as soon as NASA recovers from the Challenger disaster of 28 January 1986. Named after Edwin Hubble, who discovered the expansion of the universe, the Hubble Space Telescope will revolutionize cosmology. Though its 2.4-meter aperture is considerably smaller than the largest telescopes on Earth, it will be able to look 5 to 10 times farther into space (and consequently farther *back into time*) simply because it will not have to contend with the effects of Earth's atmosphere. The Space Telescope will be the first orbiting astronomical observatory to be used over the optical range of wavelengths. It also will be used for observations down to 1100 Å in the ultraviolet and out to 11,000 Å (1.1 microns) in the near infrared. What kind of problems can we study with it? We hope to determine an accurate value of the expansion rate of the universe. Fifty years after the expansion was discovered, the rate is still uncertain by 50 percent. We hope to elucidate the driving engines of quasars and active galaxies. We hope to determine the composition of the early universe to aid in our understanding of the universe's chemical evolution. We hope to refine our understanding—in many ways quite scanty—of the formation of stars, the enrichment of the interstellar medium, and the dynamics of galaxies and galaxy clusters. Can we find incontravertible evidence of black holes? The Space Telescope will shed light on many of these problems.

Mauna Kea in Hawaii is the site of two millimeter- and submillimeter-wavelength telescopes now under construction. They are like radio telescopes, but work at wavelengths between radio waves and the infrared. A 10-meter dish is being built by the California Institute of Technology, and a 15-meter dish is being built by the British and Dutch. These telescopes will allow us to measure the compositions and energetics of interstellar molecules and to discover new molecular species. Mauna Kea will also become the site of the 10-meter Keck optical and infrared telescope (built by Caltech and the University of California), which will have a segmented mirror of 36 sections, like a cross section of a beehive. It will have four times the light-gathering power of the 200-inch Palomar reflector.

Astronomers in Texas are planning a 7-meter ground-based reflector. Astronomers in Arizona hope to build an 8-meter reflector at Mt Graham and are working on plans for a 15-meter multiple-mirror telescope. These telescopes, which all work at optical and infrared wavelengths, underscore the Big Science aspect of modern astronomy. How it has changed since the days of Galileo's simple 'optick tube'!

There are hopes and plans for space probes as well. The Shuttle Infrared Telescope Facility will pick up where IRAS left off and will help us with such projects as finding planets orbiting other stars. The Advanced X-ray Astronomy Facility, projected to cost $500 million, will have a light-gathering power four times that of Einstein and will be used for observations of supernova remnants, quasars and other X-ray sources. A Gamma-Ray Observatory, to be launched in 1988, will bring gamma-ray astronomy to the state in which X-ray astronomy is now. The Galileo probe to Jupiter will be launched in a year or two. A Large Deployable Reflector, an orbiting 10-meter telescope for millimeter and submillimeter observations, may also fly some day. How about a manned mission to Mars?

As I write this, the first of five satellites has just passed Halley's Comet. For this once in *my* lifetime event, two Russian, two Japanese, and one ESA craft will give us the first close-ups of the nucleus of a comet. We expect to find out more about comets this year than in all of past history.

Ours is a universe filled with astonishing vistas made possible by advances in observational equipment coupled with theoretical understanding. Computer and detector technology give us many pretty pictures of members of the astrophysical zoo. Enjoy the view that we present here and contemplate what the future may bring.

Kevin Krisciunas
Hilo, Hawaii
6 March 1986

For Further Reading

Cornell, James, and Gorenstein, Paul, eds. *Astronomy from Space: Sputnik to Space Telescope*. Cambridge, Mass.: MIT Press, 1983.

Field, George B, and Chaisson, Eric J. *The Invisible Universe: Probing the Frontiers of Astrophysics*. Boston: Birkhäuser, 1985.

Gatland, Kenneth, principal author. *The Illustrated Encyclopedia of Space Technology: A comprehensive history of space exploration*. New York: Salamander Books, 1981.

Harwit, Martin. *Cosmic Discovery: The search, scope, and heritage of astronomy*. New York: Basic Books, 1981.

Henbest, Nigel, and Marten, Michael. *The New Astronomy*. Cambridge University Press, 1983.

Herrmann, Dieter B. *The History of Astronomy from Herschel to Hertzsprung*, translated and revised by Kevin Krisciunas. Cambridge University Press, 1984.

King, Henry C. *The History of the Telescope*. New York: Dover, 1955.

Sky and Telescope magazine has many articles of interest. The twice annual indices are published in the June and December issues. Look under the subject headings 'Artificial satellites and spacecraft,' 'Space and spacecraft,' or 'Observatories.'

NASA
Space
Science

The space science programs of NASA have contributed significantly to a new golden age of discovery. They have substantially advanced the frontiers of knowledge about our home planet, the relationships of the Sun and Earth and celestial phenomena.

This section opens with a brief summary highlighting new knowledge gained. More detailed presentations on knowledge acquired by analyzing data from each spacecraft follow. It should be noted that an important part of the process of discovery involves correlating data acquired by spacecraft near and far from Earth, by ground and airborne observatories, by balloons and by sounding rockets launched from a large variety of locations, and by laboratory experiments and theory development.

NASA satellites discovered the existence of the Van Allen Radiation Region around Earth. They demonstrated that Earth's magnetic field is not shaped like iron filings around a bar magnet but like a vast cosmic teardrop. It is literally blown into this shape by the solar wind, a hot electrified gas constantly speeding out from the Sun. Satellites also contributed to understanding the interaction of solar activity with the magnetic field to produce periodic radio blackouts, trip circuit breakers of electric transformers, cause magnetic compasses to become erratic and generate auroras.

NASA satellites demonstrated that the earth's tenuous upper atmosphere is not as stable and quiescent as previously believed.

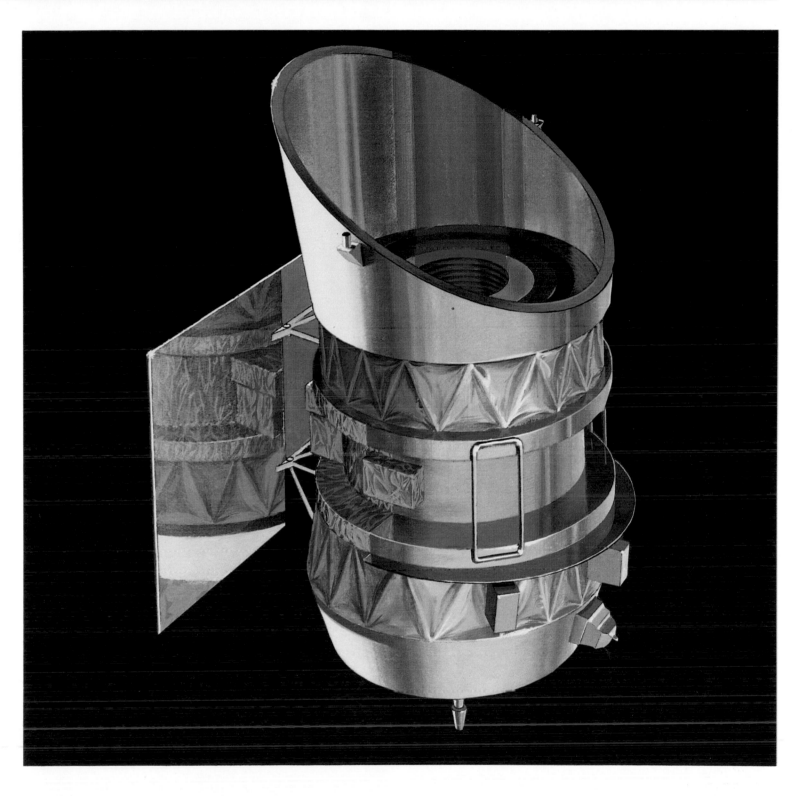

It swells by day and contracts at night. Its volume and density wax and wane with such solar events as solar flares, the 11-year solar cycle and the 27 day solar rotation period.

NASA spacecraft confirmed the existence of the solar wind. They discovered that the solar wind streams outward along the Sun's magnetic field lines. Analyzing magnetic field and solar wind data from spacecraft near and far, scientists have redrawn the picture of interplanetary space. The solar wind and solar magnetic field are now visualized as forming a vast heliosphere encompassing space billions of miles outward from the Sun.

Satellites have dramatically altered our conception of the universe. Our atmosphere blocks most of the electromagnetic radiations that can tell us about the nature of celestial objects. Satellite observa-

(*Above*) **The Infrared Astronomical Satellite (IRAS), a joint effort of the United States, Holland and the United Kingdom, was launched in January of 1983. IRAS weighed 2365 lb, was 12 ft in height with a diameter of 7 ft. Instruments included a 22-in reflector telescope and infrared equipment. The mission of IRAS was to scan the sky in search of celestial objects and previously unknown sources.**

tories viewing the heavens from above our appreciable atmosphere open a window on the universe.

Astronomers had theorized that only a small fraction of the electromagnetic radiation emitted by stars consisted of X-rays. X-rays are generated by high-energy processes, that is, among the most violent events. A sparse emission of X-rays would indicate that our universe was comparatively peaceful and slowly evolving. However, NASA satellites revealed a skyful of X-ray sources. This revolutionized the concept of the universe to one whose dynamics and evolution are governed by dramatic and enormously powerful processes. Our study of the universe draws us toward the answers to fundamental questions about the very nature of matter, life and the destiny of the stars.

Arguably, the most dramatic discoveries have resulted from observations of other planets and, if present, their satellites. The observations accentuated the uniqueness of our planet, its resources and its environment. It is a place where a variety of suitable conditions combined to create and sustain life.

Our nearest neighbor, the Moon, is a radically different world. Apollo expeditions, unmanned spacecraft and telescope observations show it to be pockmarked with huge meteorite craters partially filled with basalts from ancient lava flows. Its surface is wracked with excessive heat by day and cold by night. It is bombarded by solar radiation because it has neither an intrinsic atmosphere nor magnetic field to ward off the radiation. It has no signs of life nor even evidence that life processes have begun.

In times past, some people speculated that Venus was a twin of Earth. Venus' density, gravity and size are nearly the same as Earth's. The white clouds enveloping the planet were likened to water-based clouds of Earth. Spacecraft gave starkly different views. The beautiful white clouds are composed primarily of corrosive sulfuric acid droplets. Venus' atmosphere, 97 percent of which is carbon dioxide, has a crushing surface pressure about 100 times the Earth's at sea level. Its surface temperature is 482°C (900°F), hot enough to melt lead or zinc. Water has disappeared from the hot planet. Spacecraft confirmed that Venus' ovenlike surface temperature is due to a greenhouse effect in which Venus' mostly carbon dioxide atmosphere admits sunlight but traps outgoing heat radiation.

Mars, the planet of fabled canals, was found to have no canals but rather an extensive alluvial system from a period much earlier in its history when water perhaps flowed freely. It has an atmosphere, only about 1/100 the density of the earth's, made up mostly of carbon dioxide. Examinations of its soil reveal no signs of life. The planet's water is locked in its north polar ice cap and in a subsurface tundra. The tenuous atmosphere would turn any escaping liquid water promptly into vapor. Mars' surface is drier than driest deserts on Earth.

Studies of Mars and Venus suggest that once they had rivers and seas and that their atmospheres and temperatures were benign, but they cannot determine whether life developed during these benign periods. Spacecraft could detect on intrinsic magnetic fields around Mars or Venus. Surprisingly, a faint magnetic field was discovered around Mercury. This discovery raised questions about theories of magnetic field origin that call for rapid rotation and a liquid metal interior. Mercury rotates slowly. Its surface is pocked with ancient craters, much like the Moon's, and it is airless except for wisps of noble gases.

Close-range spacecraft observations of the giant planets Jupiter and Saturn revealed they have no solid surfaces. Beneath their deep turbulent clouds, the planets are composed mostly of liquid hydrogen. They have vast magnetic fields that apparently emanate from their rapid rotations and metallic liquid hydrogen in their interiors. The temperatures and pressures that are responsible for metallic liquid hydrogen are impossible to duplicate on Earth. Their magnetic fields are blown by the solar wind into vast teardrop shapes like Earth's and have trapped atomic particles, creating regions of

(*Right*) **A 1966 artist conception of a NASA Lunar Orbiter 28 miles above the lunar surface on a photographing mission. The four paddle-shaped solar panels provide power to the craft. This rendering has removed most of the aluminized mylar thermal barrier to show the camera equipment (twin lenses in port at lower midsection) and system electronics. The dish-like directional antenna on the far side (near the nose) is pointed toward Earth, ready to transmit photographs.**

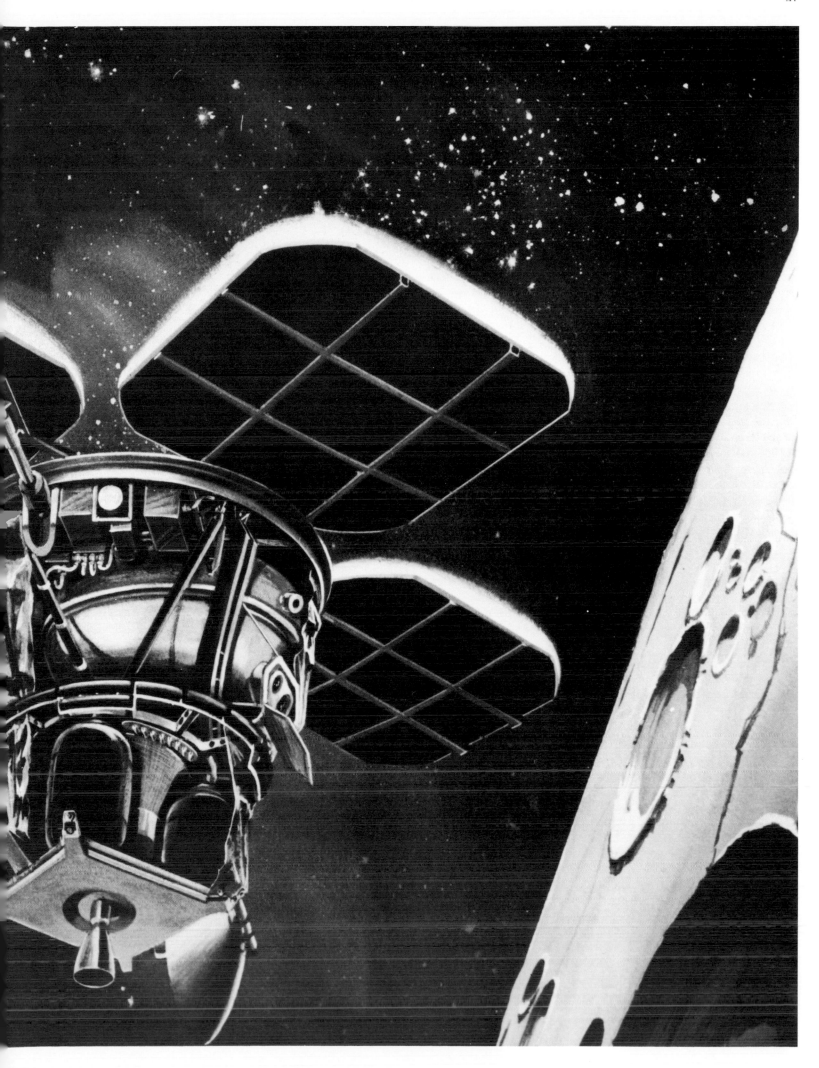

concentrated radiation far larger and more intense than Earth's.

Spacecraft data led to the discovery of many satellites around Jupiter and Saturn and a ring around Jupiter. With this ring discovery and the discovery by a NASA airborne observatory of rings around Uranus, only Neptune of the giant outer planets is not known to be ringed.

The satellites of Jupiter and Saturn appear as points of light in earth telescopes. Spacecraft sweeping nearby the planets telecast close-ups not only of the planets but also of their satellites. With these telecasts, people saw for the first time the surfaces of many of these satellites.

Jupiter has four large Galilean satellites, named for their discoverer, the Italian seventeenth-century astronomer Galileo, and a dozen smaller ones. One of the four, Ganymede, is the largest satellite in the solar system and is larger than the planets Mercury or Pluto. Its surface shows signs of progressive change before it froze solid more than 3 billion years ago. Another, Callisto, shows scars made by ancient meteorite bombardments. Its surface appears to have changed little in 4 billion years. Callisto, like Ganymede, has a surface mix of rock and water ice and is larger than the planets Mercury or Pluto. Europa, about the size of the earth's moon, has an ice-covered surface that is the smoothest seen on any celestial body. All three satellites are made up of substantial quantities of water ice.

Until viewed close up, astronomers had believed that Io, the remaining Galilean satellite, was dry, dead and rocky like our Moon. Spacecraft confirmed that Io was rocky and waterless and had active volcanoes. Spacecraft confirmed that water ice makes up most of Saturn's rings and all or a large part of Saturn's satellites. The major exception is the outermost satellite, Phoebe, which close-up pictures suggest is a captured outer solar system asteroid. Earth telescopes show vast gaps in Saturn's rings. Spacecraft close-ups reveal that the gaps are filled with many additional, very thin ringlets. Saturn's satellite Titan, larger than Mercury or Pluto, was once considered the largest satellite in the solar system. Spacecraft close-up measurements showed it was slightly smaller than Ganymede.

Titan is of interest because it is the only satellite in the solar system with a substantial atmosphere. Spacecraft studies indicate that the atmosphere is more than 8 percent nitrogen, with methane, ethane, acetylene, ethylene, hydrogen, cyanide and other organic compounds making up the rest. Its atmospheric surface pressure is about 1.6 Earth's sea-level pressure. Its atmosphere resembles that which Earth is presumed to have had in primeval times. A smog apparently caused by chemical reaction of sunlight on methane in Titan's atmosphere envelops the satellite.

Spacecraft further determined that Titan is composed of about equal amounts of rock and water. Scientists observed that with a suitable temperature Titan's environment could trigger prelife chemistry. Some speculated that the smog-generated greenhouse effect could have accumulated enough heat over the last few billion years to raise the temperature to satisfactory levels. However, Voyager 2 took Titan's surface temperature and found it was a chilling −288°F, far too cold for water to liquefy or for significant progress in prelife chemistry.

VANGUARD SATELLITES

The US Navy launched Vanguard 1 on 17 March 1958. Vanguard 1 provided information leading in 1959 to identification of the slight but geologically significant distribution called the 'pear shape' of the earth. Vanguard 2, launched 17 February 1959, transmitted the world's first picture of cloud cover from a satellite, although a wobble caused by an inadvertent bump from its launch vehicle resulted in less than satisfactory picture quality.

(Right) **A 1968 photograph of NASA's Radio Astronomy Satellite, the 38th of the Explorer series, being mated to a Delta rocket. RAE was designed to monitor low frequency radio signals in space. Sources monitored in mission 38 included the Milky Way Galaxy, possibly other galaxies, the Sun, Jupiter and the Earth environment. This Radio Astronomy Satellite was the first of two which were approved for flight under the scientific space exploration program conducted by NASA's offices of Space Science and Applications.**

Vanguard 3 was launched 18 September 1959 and contributed extensive data about the earth's magnetic field, the Van Allen Radiation Region and micrometeoroids.

EXPLORERS IN SPACE NEAR AND FAR

Since 1 February 1958, when two organizations that later became part of NASA launched Explorer 1, NASA has launched more than 60 Explorers. Explorer satellites are comparatively small and vary in size and shape. They carry a limited number of experiments into the most suitable orbits.

The Explorer designation has been assigned not only to small satellites conducting space science missions but also to those whose primary mission is in either satellite applications or technological research. Explorers devoted principally to space science programs are described in this section.

ASTRONOMY EXPLORERS

NASA's launch of Explorer 11 on 27 April 1961 was the first step of its long-range program to probe the universe's secrets that are veiled by the earth's atmosphere. Designed to monitor gamma rays, the satellite's data appeared to contradict the steady-state theory of constant destruction and creation of matter. NASA and the US Navy launched Explorers 30 and 37 on 19 November 1965 and 5 March 1968, respectively. The satellites monitored solar X-rays and ultraviolet rays during periods of declining and increasing solar activity.

Perhaps the longest satellites ever launched were NASA's Radio Astronomy Explorers 38 and 49. The tiny bodies of these satellites were each crossed with two antennas that were about three times as long as the Washington Monument is high. The antennas were unreeled to form a vast 'X' in space, where they received natural radio signals that do not ordinarily reach the earth, thus filling a gap in our radio astronomy knowledge.

Explorer 38, launched on 4 July 1968, surprised astronomers by reporting that Earth sporadically emits natural radio waves. Until then, the only planet known to emit radio waves was Jupiter. Earth was so noisy that it drowned out many other sources. However, Explorer 38 was also able to report that the Sun emitted more low-frequency radio signals than scientists anticipated. Earth's radio noise prompted NASA to make Explorer 49, the second Radio Astronomy Explorer, into a lunar-anchored satellite. After the 10 June 1973 launch, Explorer 49 was maneuvered into lunar orbit, far enough away to prevent radio interference from Earth.

Explorer 42, launched 12 December 1970 was the first of a new category of NASA spacecraft called Small Astronomy Satellites (SAS). Designed to pick up X-rays, it gathered more data in a day than sounding rockets accumulated in the nine previous years of X-ray astronomy. Atronomers used its data to prepare a comprehensive X-ray sky map and X-ray catalog. Data from Explorer 42 suggested that superclusters of galaxies may be bound together by tenuous gases whose total mass is greater than that of the optically visible galaxies. This would provide a significant percentage of the mass needed to support the theory that our expanding universe will eventually contract.

Explorer 42 was the first NASA satellite launched by a foreign nation. Italy launched Explorer 42 from its floating San Marco platform in the Indian Ocean off the coast of Africa, near Kenya. Because Explorer 42 was launched on Kenya's independence day, it was also named 'Uhuru,' which is Swahili for 'freedom.'

On 15 November 1972 Italy launched NASA's Explorer 48, the second SAS, from its San Marco platform. Explorer 48 continued the expansion of knowledge about gamma ray sources first begun by Explorer 11. It provided data that could be interpreted as supporting the theory that the universe is composed of regions of matter and antimatter. Explorer 53, the third SAS, was launched from San Marco on 7 May 1975. It discovered many additional X-ray sources, including one identified as a quasistellar object (quasar) only 783 million light years away—the closest quasar yet discovered.

Many new discoveries were made by the International Ultraviolet Explorer (IUE), launched on 26 January 1978. IUE is a joint project of NASA, the United Kingdom and the European Space Agency. IUE data supported a theory that a black hole with the mass of a thousand solar systems existed at the center of our Milky Way galaxy and revealed that our galaxy had a halo of hot gases. The data provided evidence that so-called twin quasars were actually a double image of the same object. Light waves from the quasar are bent around a massive elliptical galaxy that acts as a gravitational lens to produce the double image picked up by ground observatories.

ATMOSPHERE EXPLORERS

Atmosphere Explorers have confirmed or redrawn our conceptions of the earth's tenuous upper atmospere. The first, Explorer 8, was launched 3 November 1960. It confirmed that temperatures of electrons in the upper ionosphere are higher by day than by night. It discovered that oxygen predominates in the ionosphere up to an altitude of about 650 miles where helium predominates. A secondary experiment indicated that micrometeoroid quantities varied inversely with size.

Air Density Explorers 9, 24 and 39 were launched 16 February 1961; 21 November 1964; and 8 August 1968, respectively. These were essentially 12-foot balloons of aluminum foil and plastic laminate that were inflated in orbit. Air drag on the satellite indicated air density. The satellites revealed that atmospheric density varied from day to night, with the 27-day rotation period of the Sun, with the 11-year solar cycle and with violent eruptions on the Sun.

Explorer 25 was launched on the same Scout booster that orbited Explorer 24, the first multiple launch by a single vehicle, while Explorer 40 was launched with the same Scout vehicle as Explorer 39. Explorers 25 and 40, called Injuns and University Explorers because they were built by the University of Iowa, demonstrated a correlation between air density and solar radiation. Explorers 17 and 32, launched 2 April 1963 and 25 May 1966, gathered information about the composition of neutral atoms and molecules. Explorer 17 confirmed Explorer 8 indications of a belt of neutral helium in the upper atmosphere.

In radio-echo soundings of the ionosphere, radio signals at different frequencies were transmitted from the ground. The reflected frequency disclosed electron density; the return time indicated the altitude or distance at which the density was encountered. Ground-based sounding cannot provide information about the upper ionosphere because electron density increases up to a certain altitude and then tapers off. In addition, many areas of Earth are too remote or inaccessible for ground-based radio sounding. These problems were solved by using satellites as topside sounders. They beamed radio waves into the ionosphere from altitudes far above the region of maximum electron density.

NASA topside sounders were Explorer 20, launched 25 August 1964 and Explorer 31, launched 28 November 1965. Scientists correlated data from these satellites with data from the Canadian topside sounders Alouette 1 and 2. Explorer 22, an ionosphere beacon launched 10 October 1964, also measured electron density in the ionosphere. It was built with quartz reflectors for the first major experiment in laser tracking.

The thermosphere, a region of the upper atmosphere, was believed to be relatively stable until Explorers 51, 54 and 55 were launched. These satellites were equipped with onboard propulsion systems that enabled them to dip deep into the atmosphere and pull out again, taking measurements and providing extensive data about the upper thermosphere. They found that the thermosphere behaved unpredictably with winds 10 times stronger than normally found at the earth's surface. They discovered abrupt and constantly changing wind shears. Their data contributed significantly to knowledge about energy transfer mechanisms and photochemical processes (such as those that create the ozone layer) in the atmosphere. The launch dates for these three Explorers were 16 December 1973, 6 October 1975 and 19 November 1975.

The Solar Mesosphere Explorer, launched 6 October 1981, provided comprehensive data on how solar radiation creates and destroys ozone in the mesosphere, an atmospheric layer below the thermosphere and above the stratosphere. The University of Colorado designed and built the Solar Mesosphere Explorer and operated it for a year after launch.

GEOPHYSICAL EXPLORERS

NASA's first successful satellite launch was Explorer 6, orbited 7 August 1959. Explorer 6 added to information about the Van Allen Region and micrometeoroids, and also telecast a crude image of the north Pacific Ocean. Explorer 7, launched 13 October 1959, provided data revealing that the Van Allen Region fluctuated in volume intensity and suggested a relationship of the region with solar activity. It indicated that variations in solar activity may also be related to the abundance of cosmic radiation in the earth's vicinity, magnetic storms and ionospheric disturbances.

Explorer 10, launched on 25 March 1961 to gather magnetic field data, was the first spacecraft to obtain information that suggested that the interplanetary magnetic field may actually be an extension of the Sun's field carried outward by the solar wind.

With Explorer 12, launched 15 August 1961, scientists were able to reach many conclusions about space: the Van Allen Region is a single system of charged particles rather than several belts; the earth's magnetic field has a distinct boundary; the solar wind compresses the earth's magnetic field on the Sun's side and blows it out on the other; and geomagnetic storms that cause radio blackouts and power outages may result from solar flares.

On 2 October 1962 NASA launched Explorer 14 to monitor the Van Allen Radiation Region during a period of declining solar activity. Scientists in the meantime discovered that the United States project Starfish, involving a high-altitude nuclear burst in July 1962, had created another artificial radiation belt. Explorer 14 was joined by Explorer 15 on 27 October to help monitor this belt. The two satellites' data helped ease scientific anxiety by indicating that atomic particles making up the belt were rapidly decaying.

NASA launched Explorer 26 on 21 December 1964, and its data increased understanding of how atomic particles traveling toward Earth from outer space are trapped by the earth's magnetic field

(*Above*) **A 1973 photograph of the Atmosphere Explorer-C. This 1450-lb spacecraft carried 14 scientific experiments and was launched into an extremely egg-shaped orbit designed to acquire data on the composition of the thermosphere of Earth.**

and how they spiral inward, along the earth magnetic field lines in the northern and southern latitudes, interacting with the atmosphere to generate auroras.

Explorer 45, launched 15 November 1971, further investigated the relationships of geomagnetic storms, particles radiation and auroras. It was the second satellite launched by an Italian crew from the San Marco platform off Kenya in the Indian Ocean.

NASA's Interplanetary Monitoring Platforms, or IMP Explorers, added significantly to knowledge about how the earth's magnetic field and the Van Allen Radiation Region fluctuate during the 11-year cycle. IMP Explorer 18 confirmed that the earth's magnetic field was shaped like a giant cosmic teardrop. It discovered a shockwave ahead of the earth's field, caused by the impact of the speeding solar wind with the earth's field. Between the shockwave and the magnetopause, or magnetic field boundary, Explorer 18 discovered a turbulent region of magnetic fields and atomic particles.

IMP Explorer 33 was the first satellite to provide evidence that the geomagnetic field on the earth's night side extends beyond the Moon. IMP Explorer 35, a lunar-anchored (lunar orbiting) IMP, gathered data about micrometeoroids, magnetic fields, the solar wind and radiation at lunar altitudes. Its instruments revealed the

Moon to be what one scientist termed a 'cold nonmagnetic nonconducting sphere.'

A two-satellite Dynamic Explorer project, launched simultaneously on 3 August 1981, significantly contributed to data on coupling of energy, electric currents, electric fields and plasmas (hot electrified gases) between the earth's magnetic field, the ionosphere and the rest of the atmosphere. Among their discoveries were nitrogen ions in the geomagnetosphere. They also confirmed existence of the polar wind, which is an upward flow of ions from the polar ionosphere.

The Dynamic Explorers complemented studies of the three International Sun-Earth Explorers (ISEE), a joint project of NASA and the European Space Agency. ISEE 1 and 2 were launched into earth orbit on 22 October 1977. ISEE 3 was placed in a heliocentric orbit near the Sun-Earth libration point.

The ISEE program focused on solar-terrestrial relationships as a contribution to the International Magnetospheric Study. The three spacecraft obtained a treasure trove of new information on the dynamics of the geomagnetosphere, the transfer of energy from the solar wind and energization of plasma in the geomagnetotail. For example, ISEE 1 found ions from our ionosphere accelerated in the geomagnetotail to fairly high energies. Previously, scientists thought these high-energy particles originated from the solar wind.

In 1982, when its mission was completed with fuel to spare and all instruments in working condition, ISEE 3 was put through a series of complex maneuvers to explore the earth's magnetotail through December 1983, and to fly across and study the wake of Comet Giacobini-Zinner in September 1985.

ASTRONOMICAL OBSERVATORIES

Man's views of the universe is narrowly circumscribed by the atmosphere, which blocks or distorts most kinds of electromagnetic radiations from space. Analyses of these radiations (radio, infrared, visible light, ultraviolet, X-rays and gamma rays) give important new information about the phenomena in our universe.

The small Astronomical Explorers indicated the great potential for acquiring new knowledge by placing instruments above the earth's obscuring atmosphere. Consequently, NASA orbited a series of large astronomical observatories bearing a great variety of instruments which have significantly widened our window on the universe.

The first successful large-scale observatory was Orbiting Astronomical Observatory 2 (OAO 2), nicknamed *Stargazer*, which was launched 7 December 1968. In its first 30 days, OAO 2 collected more than 20 times the celestial ultraviolet data acquired in the previous 15 years of sounding rocket launches.

Among the volumes of data provided by OAO 2 were numerous discoveries. It learned that stars that are many times more massive than our Sun are hotter and consume their hydrogen fuel faster than estimated on the basis of ground observations. OAO 2 data contributed to resolving a disparity between observations made from the ground and theories of stellar evolution. Another stellar theory was brought into question by *Stargazer*. According to this theory, the intensity of celestial objects should be less in ultraviolet light than in visible light. However, several galaxies that looked dim in visible observations from the earth were bright in ultraviolet observations by *Stargazer*.

(*Left*) An artist conception of the ISEE-3 spacecraft at just 72 miles above the lunar surface in December 1983, enroute to an encounter with Comet Giacobini-Zinner in September of 1985. The spacecraft flew close to the moon in order to use the gravity of the Earth's satellite to provide additional thrust by catapulting it toward the comet rendezvous point.

OAO 3 (Copernicus)

Spacecraft Description—A 4,900-pound, octagonal cylinder 7 feet wide and 10 feet long. Two paddle-like solar arrays gave it an overall width of 21 feet when they were deployed. It was three-axis stabilized with a pointing accuracy of less than 0.1 second of arc—the most precise ever achieved by a spacecraft. More than one-fifth (1000 pounds) of the total weight was scientific payload.

Project Objectives—To study interstellar absorption of hydrogen, oxygen, carbon, silicon and other common elements in the interstellar gas, and to investigate ultraviolet radiation emitted from so-called young hot stars in wavelength regions between 930 and 3,000 Å.

Spacecraft Payload—The Princeton Experiment Package (PEP), in the central experiment tube, included a 10-foot-long ultraviolet telescope, a 32-inch mirror, an ultraviolet spectrometer and sensors for the telescope guidance system, able to view stars as faint as seventh magnitude. Scientific objectives of the PEP included study of the abundance and temperature distribution of interstellar gas and study of the structure of stellar atmospheres of young hot stars. The University College of London Experiment Package, mounted in an upper bay of the main body, consisted of three small telescopes and a collimated proportional counter. It permitted pinpointing many recently discovered X-ray sources more precisely.

Project Results—Launched 21 August 1972 from Kennedy Space Center. Preliminary results from the PEP included detection of large quantities (more than 10 percent) of molecular hydrogen in the denser interstellar dust clouds; (hydrogen also occurs in atomic form in these regions); observation of surprisingly large amounts of deuterium—heavy hydrogen—in interstellar dust clouds; (deuterium is a basic element for fusion in the formation of stars, and current theories suggest that much of it should already have been used up. These theories may have to be revised in view of the abundance of deuterium observed); determination that lesser amounts of heavier elements exist in clouds than in the Sun; determination that some solid particles or dust grains in interstellar clouds are smaller than believed previously—some less than one-millionth of an inch in diameter. Preliminary results from the University College of London Experiment Package indicate that the period of rotation of the Cygnus X-1 binary system is increasing at a rate perceptible after the first month of operations.

Stargazer was able to observe Nova Serpentis in 1970 for 60 days after its outburst. It confirmed that mass loss by the nova was consistent with theory. *Stargazer* observations of the Comet Tago-Sato-Kosaka supported the theory that hydrogen is a major constituent of comets. It detected a hydrogen cloud as large as the Sun around the comet. Because of our atmosphere, this hydrogen cloud could not be detected by ground observatories. Looking toward the earth, *Stargazer* also reported that the hydrogen in the earth's outer atmosphere is thicker and covers a larger volume than previous measurements indicated.

Launched 21 August 1972, OAO 3 was named for the famed Polish astronomer Copernicus. The satellite provided much new

OAO 1

Spacecraft Description—Octagonal aluminum cylinder 10 feet long and 7 feet wide. Experiments housed in 4-foot central cylinder running length of the body. Sunshade mounted at top of experiment tube shielded instruments from direct solar rays. Two solar arrays of three panels each attached to opposite sides of spacecraft, giving OAO 1 a wingspread of 21 feet; two 9.5-foot balance weight booms also attached. Weight: 3900 pounds.

Project Objectives—OAO 1 carried four experiments to study, for the first time over an extended period, the UV, X-ray and gamma ray regions from above the obscuring and distorting effects of the earth's atmosphere.

Spacecraft Payload—Experiment instrumentation weighed 1000 pounds and included: 1) seven UV telescopes to view about 200 selected stars and nebulas in 1100 to 3000 A region, 2) high-energy gamma ray detector similar to device carried on Explorer XI, 3) gas proportional counter to define and map X-ray sources and 4) low-energy gamma ray detector to survey photon sources in 2 to 180 K range.

Project Results—OAO 1 was launched 8 April 1966 from Kennedy Space Center and placed in near-circular orbit of 496 to 502 miles at 35° inclination. Initial stabilization was achieved but on its second day in orbit the satellite's primary battery began to overheat and soon malfunctioned, rendering OAO 1 inoperative.

OAO 2 *(pictured at left)*

Spacecraft Description—Octagonal aluminum cylinder 10 feet long and 7 feet across, with experiments mounted in 4-foot inner cylinder running length of body, trapdoor-like sun shutter attached to top for shielding experiments from direct solar rays. With solar arrays unfolded on each side, wingspread reached 21 feet. Other appendages included two 9.5-foot balance booms extending outward, one from each side at top. Weight: 4446 pounds.

Project Objectives—Observe young, hot stars in ultraviolet spectrum; observe interstellar gas; survey and produce pictorial maps of over 700 stars daily—all from the vantage point of a precisely stabilized platform above the obscuring and distorting effects of the earth's atmosphere.

Spacecraft Payload—Two groups of experiments: one, a 450-pound package, contained seven telescopes to make spectrophotometric measurements in ultraviolet between 1000 and 3000 Å; second experiment group weighed about 500 pounds, employed four large-aperture TV cameras with broadband photometers to scan between 1050 and 3000 Å. Power system featured 10 percent greater solar cell area than on OAO 1, addition of undervoltage detector, and a change from sequential to parallel charging of batteries to avoid overcharging as occurred on OAO 1.

Project Results—OAO 2 was launched from Kennedy Space Center on 7 December 1968 into a near-nominal 479/485-mile orbit at 34.997°.

High Energy Astronomy Observatory-A (HEAO 1)

Spacecraft Description—A 10.5-foot-high hexagonal experiment module atop HEAO's octagonal spacecraft equipment module. A rectangular solar array deployed from the top side of the experiment module, bringing total spacecraft height to 20 feet. Two other arrays affixed along one side of the experiment module combined with the topside array to deliver 460 watts of electrical power to HEAO's experiments and spacecraft subsystems. HEAO 1 was one of the heaviest satellites ever flown, weighing in at nearly 7,000 pounds. (Of that, more than 2,600 pounds was scientific payload weight.) HEAO 1's 'spin' could be arrested for several hours, enabling its use as a stationary pointing platform to study items of special interest. HEAO's communications system was dual-frequency S-band with onboard tape recorders allowing 220 minutes of data storage.

Project Objectives—Survey and map X-ray sources throughout the celestial sphere; measure low-energy gamma-ray flux; investigate astronomical phenomena such as black holes, quasars, and pulsars; provide an all-sky survey of 'soft' X-rays (0.1 keV) through high-energy (10 MeV) gamma rays, with accurate location of their sources.

Spacecraft Payload—Four science instruments: large area X-ray survey experiment; cosmic X-ray experiment; scanning modulator collimator experiment; and the hard X-ray and low gamma ray instrument.

Project Results—HEAO 1 was launched 12 August 1977 from Kennedy Space Center and placed in a circular near-Earth orbit of about 240 miles with an inclination of about 22 degrees.

HEAO 2

Spacecraft Description—Two main parts: an experiment module and a spacecraft equipment module (SEM). The experiment module was eight-sided, about 10 inches wide and 22.4 inches high from its attachment to the SEM to the top of the solar panels. The SEM was an octagon-shaped prism, 3.2 inches high and 9 inches in diameter. The spacecraft had an overall height of about 22 feet; at launch it weighed about 6937 pounds.

Project Objectives—Carrying the largest X-ray telescope ever built, HEAO 2 was the second in a series of three large observatories designed to study the 'high energy universe,' i.e., X-rays, gamma rays and cosmic particles. The scientific objectives of the HEAO program were to learn more about some of the most puzzling objects in the universe: pulsars, neutron stars, black holes, quasars, radio galaxies and supernovas. Many of these radiate only in the X-ray, gamma ray and cosmic ray regions of the electromagnetic spectrum and cannot be studied from the ground because of the obscuring effects of the earth's atmosphere.

Spacecraft Payload—The X-ray telescope had a 23-inch-wide mirror and a focal length of 11 feet. Images acquired by the telescope were transmitted to ground stations. The experiment module also contained five other scientific instruments, four of which shared the use of the telescope to make a variety of measurements of X-rays emitted by stellar objects. The fifth instrument, independent of the telescope, measured properties of X-rays beyond the telescope's energy range.

Project Results—Launched from Kennedy Space Center on 13 November 1978. The spacecraft was injected into a near-circular orbit.

data on star temperatures, chemical compositions and other properties. It continued studies of the outer atmospheres of Earth, Mars, Jupiter and Saturn. It gathered data on the black-hole candidate Cygnus X-1, so named because it is the first X-ray source discovered in the constellation Cygnus. Much of its data supported the hypothesis that Cygnus X-1 is a black hole. A black hole is a one-time massive star that has collapsed to such density that it does not permit even light or other electromagnetic radiations to escape it. Scientists can study Cygnus X-1 because it is part of a binary star system and has a visible companion. In addition, according to theory, a substantial part of the matter dragged into a black hole is transformed into X-rays and gamma rays that are radiated into space before they reach the point of no return.

Copernicus also observed that interstellar dust clouds have fewer heavy elements than our Sun. This supported the contention that the Sun and planets coalesced from the debris of an ancient supernova. The spacecraft made two findings suggesting that star formation may be common. The first was that larger amounts of hydrogen molecules than expected existed in interstellar dust clouds. Secondly, surprisingly large amounts of deuterium (an element with the same atomic number and the same position in the Table of Elements as hydrogen but with twice the mass) were also detected in interstellar dust clouds. This contradicted a theory that most deuterium, a basic element for atomic fusion in stars, has been exhausted.

High Energy Astronomy Observatory 3 (HEAO 3)

Spacecraft Description—A boxlike experiment module atop a hexagonal spacecraft equipment module. Two large solar arrays and one smaller solar array were attached to one side of the experiment module. They provided HEAO 3 with 415 watts of electrical power. A conical omnidirectional antenna was atop the spacecraft. Weight: 6390 pounds.

Project Objectives—HEAO 3 carried two cosmic ray experiments and one gamma ray spectrometer to gather information on the origin, propagation, and acceleration mechanism for cosmic rays observed across the far reaches of space.

Spacecraft Payload—Three instruments: a high-resolution gamma ray spectrometer, a cosmic ray experiment, and a heavy nuclei experiment.

Two views of the Crab Nebula, the remnant of a super-nova that exploded in 1054 AD, a reflector telescope photograph (*bottom*) and an X-ray image (*top*). The bright object in the center of the X-ray is a pulsar which spins 30 times per second and emits regular signals, once per revolution. This X-ray also shows a very complex transfer of energy from the pulsar to the surrounding medium.

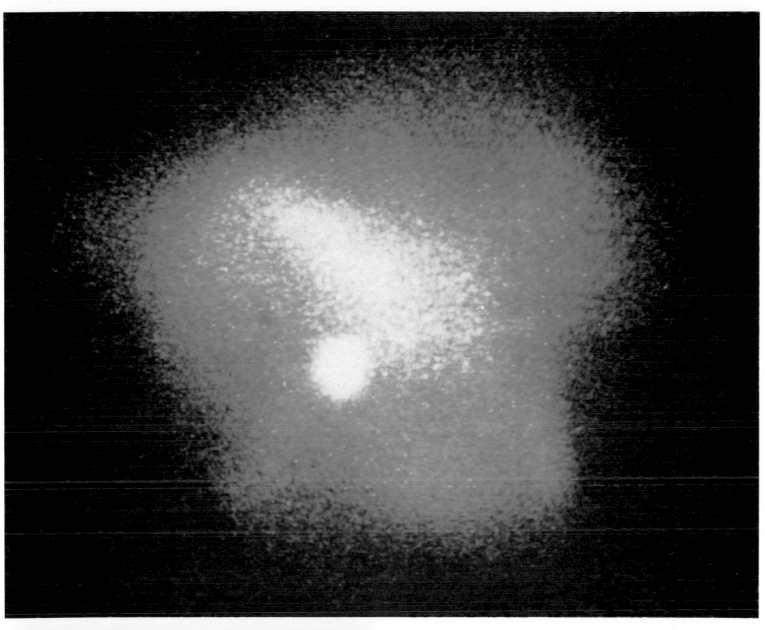

HIGH ENERGY ASTRONOMY OBSERVATORIES (HEAO)

Three High Energy Astronomy Observatories portray a universe in constant turbulence with components repeatedly torn apart and recombined by violent events. HEAO 1, launched 12 August 1977, also discovered a new black-hole candidate near the constellation Scorpius, bringing the total to four. Other black-hole candidates are in or near the constellations Cygnus, Circinus and Hercules. Another major result of HEAO 1 was the superhot superbubble of gas 1200 light years in diameter and about 6000 light years from Earth. Centered in the constellation Cygnus, the bubble has enough gas to create 10,000 suns. HEAO 1 also raised the catalog of X-ray sources from 350 to about 1500.

The *Einstein* observatory, nicknamed for the famous mathematician, is HEAO 2, launched 13 November 1978. *Einstein* was equipped with more sensitive instruments than HEAO 1. Thus, it was able to discern that the X-ray background observed by HEAO 1 was not coming from diffuse hot plasmas but from quasars. *Einstein* also provided the first pictures of an X-ray burster that is apparently located at the center of a globular cluster called Terzan 2. The bursters are frequently associated with clusters of old stars, and are usually explained in terms of gases interacting violently with neutron stars or black holes, emitting very short bursts of X-rays. *Einstein* also returned data on X-ray spectra of

supernova remnants that support the theory that our system was formed from debris of an ancient supernova.

Whereas HEAO 1 and 2 X-ray measurements related principally to atomic interactions and plasma processes associated with stellar phenomena, HEAO 3, launched 20 September 1979, scanned the universe for cosmic ray particles and gamma radiation. The events that HEAO 3 measures result from nuclear reactions in the hearts of stellar objects and the elements they create. HEAO 3 has observed in the Milky Way's central region gamma rays that apparently emanate from the annihilation of electrons and positrons (the antimatter equivalent of electrons). It has also discovered an object emitting energy in the form of gamma rays equivalent to 50,000 times the Sun's total output. The object is 15,000 light years from Earth and appears to be undergoing processes on a comparatively small scale that are believed to occur in quasars on a large scale.

The Netherlands Astronomical Satellite (NAS), a cooperative program of NASA and the Netherlands, launched 30 August 1974, was a small X-ray and ultraviolet orbiting observatory. Among discoveries from its data are X-ray bursters, sources that emit bursts of X-rays for seconds at a time.

INFRARED ASTRONOMY SATELLITE (IRAS)

IRAS is a joint project of NASA, the Netherlands and United Kingdom. Launched on 25 January 1983, it revealed many infrared sources in the Large Magellanic Cloud, 155,000 light years from Earth, that are not visible from Earth, helping scientists compile the first catalog of infrared sky sources. Because all objects, even

IRAS was a tri-nation effort. Built in the Netherlands the spacecraft (*below*) is shown being readied for flight at the Fokker BV plant. The telescope (*above*) was built by the United States. IRAS was launched in January of 1983 (*facing page*). The mission of the IRAS was to scan the entire sky in search of infrared radiation from celestial objects and previously unknown sources. The Earth atmosphere absorbs much of the infrared radiation from space and is itself a strong source of radiation. As a result, high sensitivity infrared observations can only be made from space.

continued on page 57

The Andromeda Galaxy viewed through the 200″ optical telescope at Mount Palomar in Southern California.

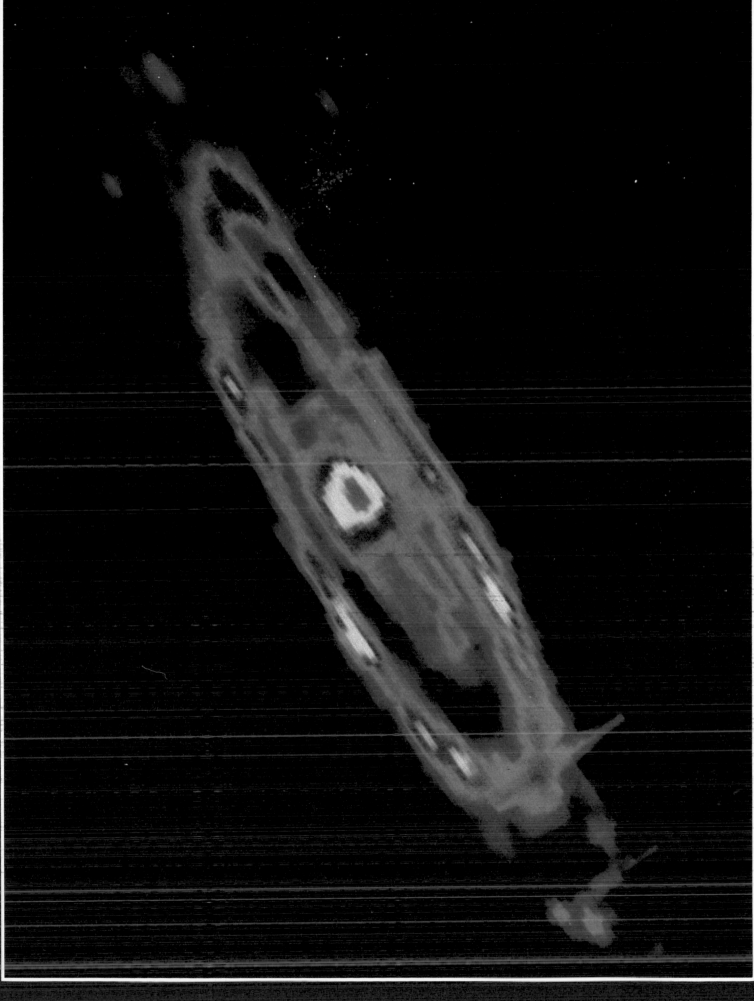

The Andromeda Galaxy imaged in brilliant color by the infrared sensors of IRAS.

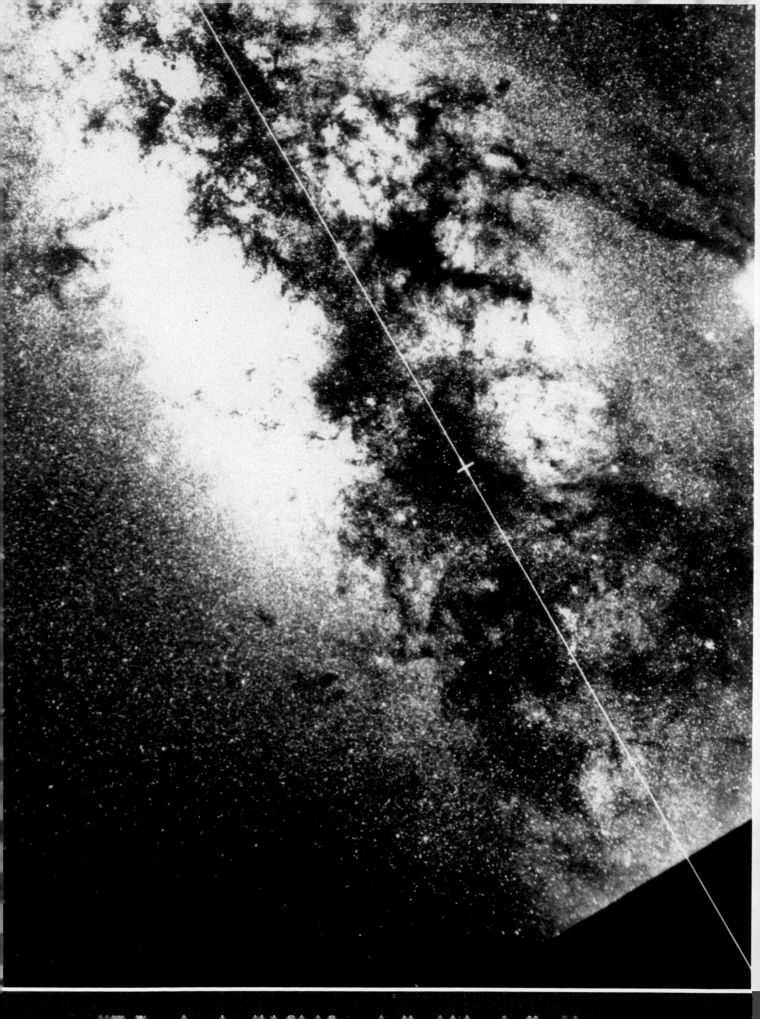

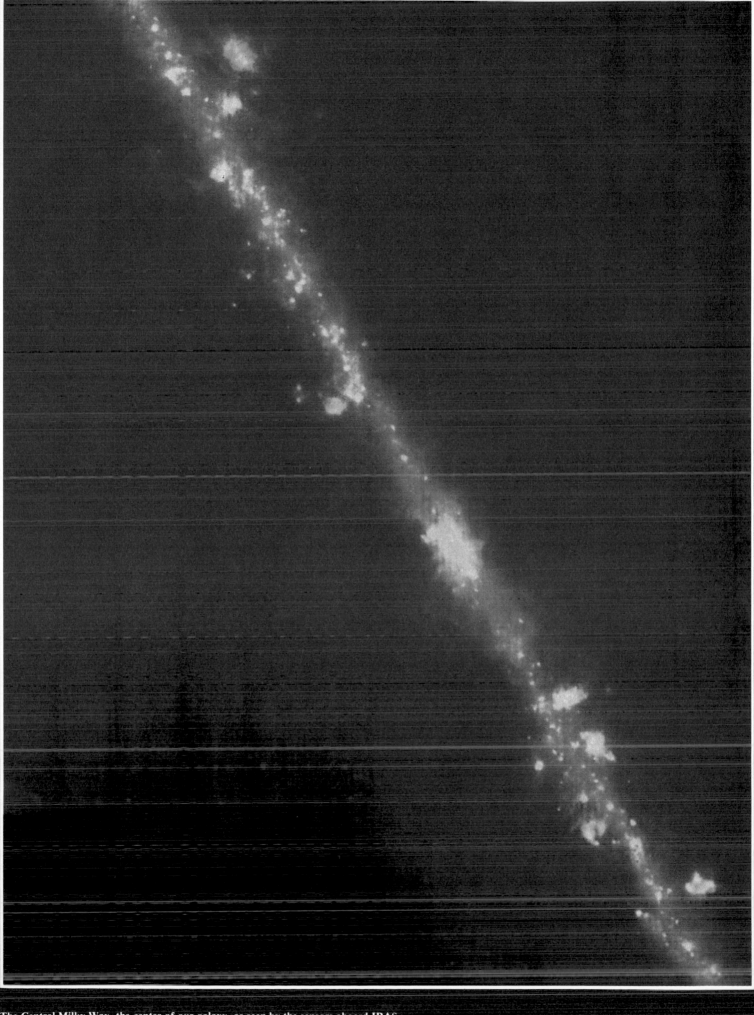

The Central Milky Way, the center of our galaxy, as seen by the sensors aboard IRAS.

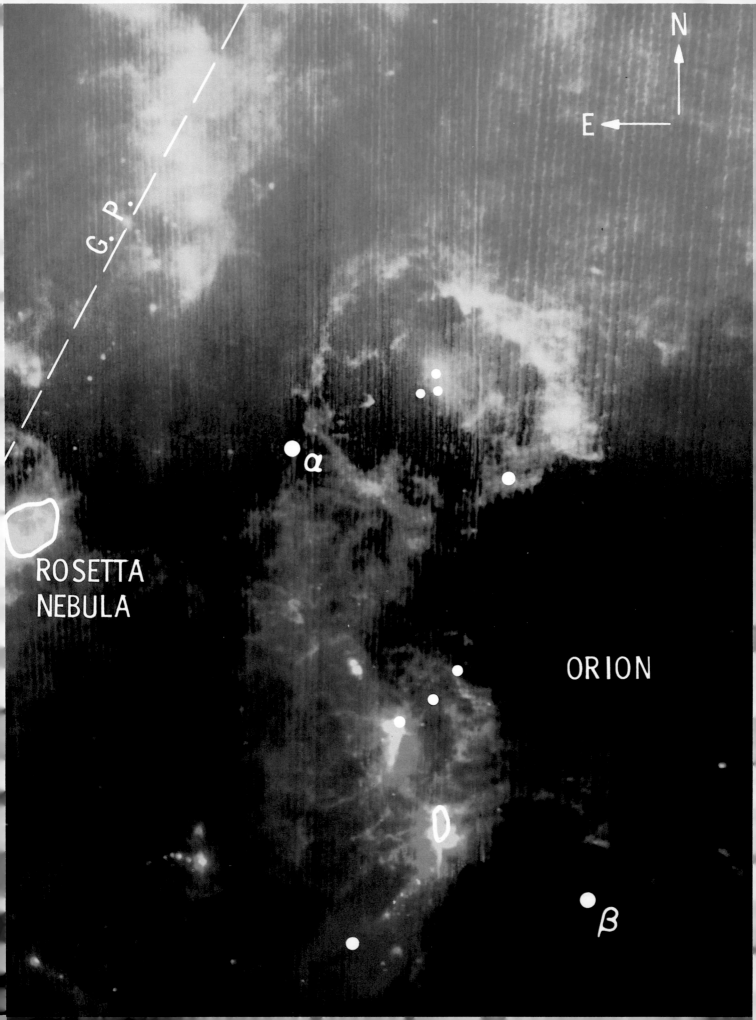

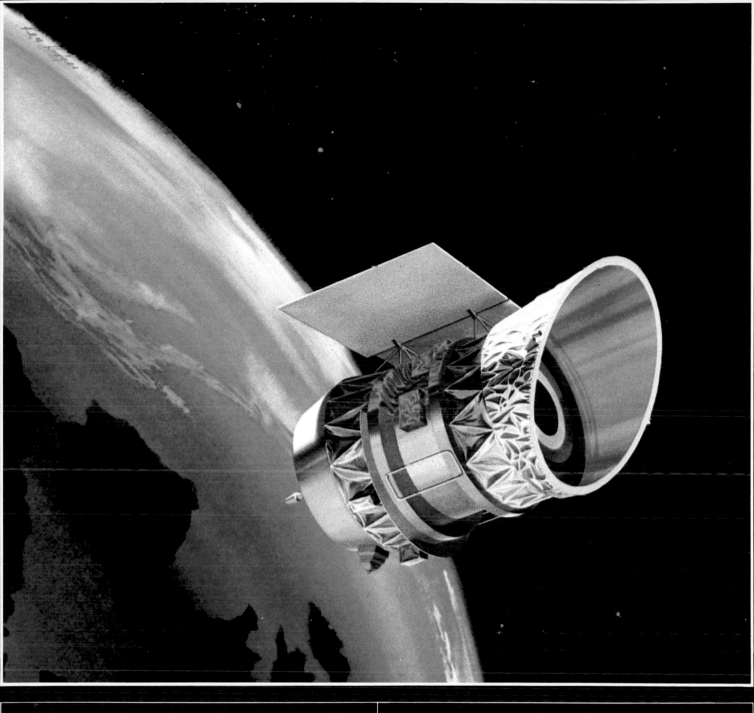

Infrared Astronomical Satellite (IRAS)

The mission of the Infrared Astronomical Satellite (IRAS) was to scan the entire sky in search of infrared radiation from galaxies, dust clouds, stars, solar system objects, and previously unknown sources. A new map of the infrared universe will be produced from IRAS data.

Earth's atmosphere absorbs much of the infrared radiation from space and is itself a strong source of radiation. Therefore, high-sensitivity infrared observations can only be made from a telescope operating above the atmosphere. The IRAS telescope was cooled to within a few degrees of absolute zero by liquid helium and scanned the sky from an orbit 500 miles high.

IRAS detected cool objects that emit the bulk of their radiation in the infrared, but so little radiation in the visible that they cannot be seen by even the most powerful optical telescope. Because infrared radiation passes freely through interstellar dust clouds, many objects that are hidden from the view of optical telescopes were clearly observed by IRAS.

IRAS was launched from the NASA Western Test Range in California on 25 January 1983. The mission ended on 21 November 1983 when the telescope's helium was depleted. IRAS was a joint project of the United States, the United Kingdom and the Netherlands. The Jet Propulsion Laboratory manages the project for the United States.

The Orion Nebula (*left*) as imaged by the Infrared Astronomical Satellite (IRAS) (*above*). Nebulas are diffuse masses of interstellar dust or gas or both. IRAS was specially designed to detect objects which emitted radiation in the infrared. The mission of this satellite was to scan the entire sky in search of infrared sources in the universe. IRAS, a joint effort of the United States, the United Kingdom and the Netherlands, was launched from the Western Range in California in January of 1983. The telescope was cooled to within a few degrees of absolute zero by liquid helium and scanned the heavens from an orbit five hundred miles above Earth.

cool dark ones that may be the black cinders of dead stars, radiate infrared light, IRAS may discover many other invisible objects. It has revealed stars being born in thick opaque clouds of gas. In April 1983 IRAS detected a new comet, which came within 3 million miles of Earth in May, the closest comet approach in 200 years. Most recently it discovered a possible new solar system near the star Vega.

ORBITING SOLAR OBSERVATORIES (OSO)

Using X-ray and ultraviolet sensors, OSO program acquired a rich harvest of solar data during the 11-year solar cycle when solar activity went from low to high and then back to low. The satellites photographed for the first time the birth of a solar flare, a great outburst of matter and energy from the sun. When directed toward Earth, solar flares can cause blackouts of communications and electricity, force magnetic compasses to spin crazily, and enhance displays of the northern and southern lights. OSOs also discovered evidence of gamma radiation resulting from solar flares, indicating nuclear reaction in the flares.

OSO 1 showed a correlation between fluctuations in temperatures of the earth's upper atmosphere and variations in solar ultraviolet ray emissions. OSO 3 revealed that the center of our galaxy was the source of intense gamma radiation, which, when confirmed by HEAO 3, led to speculation that matter and anti-matter were annihilating each other, leaving energy in the form of gamma rays. OSO 5 data revealed that the earth's upper atmosphere may contain as much as 10 times the amount of deuterium (a form of hydrogen with twice its mass) previously estimated.

OSOs discovered and provided information on solar poles, where there were cooler and thinner gases than in the rest of the corona. Later, they discovered comparable phenomena on other areas of the Sun, and scientists named them solar holes. OSO observed and reported on the dramatic coronal transient, a solar explosion hurling out hundreds of thousands of tons of material in huge loops at millions of miles per hour.

The launch dates of the eight OSOs are as follows:

OSO 1 7 March 1982   OSO 5 22 January 1969
OSO 2 3 February 1965   OSO 6 9 August 1969
OSO 3 8 March 1967   OSO 7 29 September 1971
OSO 4 18 October 1967   OSO 8 21 June 1975

SOLAR MAXIMUM MISSION (SMM)

Infinitesimal (.001) reductions in solar energy output that may be related to unusually harsh winters and cool summers on earth were discovered by the SMM satellite. SMM was launched on 14 February 1980 to study the Sun during the high part of the solar cycle. SMM also made the first clear observations of neutrons traveling from the Sun to the earth after a flare and confirmed that fusion, the basic process that powers the Sun, occurs in the solar corona during a flare. A malfunction in January 1981 caused the spacecraft to lose altitude control. It was retrieved by the manipulator arm of the Space Shuttle *Challenger* in April 1984 and successfully repaired by astronauts George Nelson and James van Hoften. The SMM was rereleased into space on 12 April to continue observations of the Sun and later Halley's Comet.

ORBITING GEOPHYSICAL OBSERVATORIES

A half dozen OGOs have provided more than a million hours of scientific data from about 130 different experiments relating to the earth's space environment and Sun-Earth interrelationships.

NASA Orbiting Geophysical Observatory satellites (OGO) (*above*) being readied for flight. The OGO satellites, launched from the Kennedy Space Center, studied the Earth's space environment in order to learn more about the Earth-Sun relationship and about the Earth as a planet. The Orbiting Solar Observatory satellite (OSO-8) (*facing page*) during testing at Hughes Aircraft Company's El Segundo, California space and communications facility. OSO spectrometers, built in France and at the University of Colorado, made fine structure studies of the chromosphere and took high resolution ultraviolet spectrometer measurements.

Solar Maximum Mission (SMM)

Spacecraft Description—The solar maximum mission spacecraft 'Solar Max' was approximately 13 feet in length, feet in diameter, and modular in design. The upper portion of the spacecraft was the instrument module, which housed all solar observation instruments and the fine-pointing sun sensor system for aiming control.

Below the instrument module was the multimission modular spacecraft (MMS), a 5-foot triangular framework, which housed the essential attitude control, power, communications and data handling systems. Two fixed solar paddles were attached to a transition adaptor between the upper instrument module and the lower spacecraft bus. The paddles supplied power to the spacecraft during the daylight portion of orbits while three rechargeable batteries supplied power at night.

Program Objectives—Scientists have long been trying to understand the many complex phenomena that combine to form a 'flare'. A wealth of individual studies of the different manifestations of the flare event exists and these investigations have contributed significantly to the development of a conceptual picture of the processes that may be taking place. There is, however, still considerable uncertainty as to the interrelationship between these processes. The resolution of the many questions requires temporally and spatially coordinated observations of a flare over the wide range of its complex phenomena.

The Solar Maximum Mission (SMM) was designed to provide these observations. The spacecraft was launched in early 1980 near the peak of maximum solar activity and carried a payload of six instruments specifically selected to study the major short wavelength and coronal manifestations of flares.

The payload was to obtain data on the storage and release of flare plasma and mass ejection. Complementary ground-based observations of the flares' radio and optical emissions were to be made as part of the SMM guest investigator program and coordinated in situ measurements of the flare particle emissions to be obtained by the International Sun-Earth Explorer C satellite. By organizing such a comprehensive program at a time in the solar cycle when activity was to be at a peak, it would be possible to study all aspects of a wide variety of different types of active phenomena including active regions, flares and transient events, and hence to answer many questions surrounding these phenomena.

Spacecraft Payload—The observatory was approximately 13 feet in length and encapsulated into a circular envelope 7.5 feet in diameter. The construction was modular, providing the capability for in-orbit change out of the spacecraft subsystem modules. The upper portion of the spacecraft, 7.5 feet, comprised the instrument module that housed all the solar payload instruments as well as the fine pointing sun sensor system for pointing control. Below the instrument module, and separated from it by a transition adaptor, was the multimission modular spacecraft, a triangular framework supporting three modules that housed the essential components of the three spacecraft subsystems: attitude control, power, and communication and data handling. The transition adaptor supported the two fixed solar paddles, which

supplied between 3000 and 1500 watts of power. The spacecraft structure also supported two other components, the signal conditioning and control unit, and the high-gain telemetry antenna system. Weight: 5104 pounds.

The spacecraft carried the following scientific instruments:

- Gamma ray spectrometer—to measure the intensity, energy, and Doppler shift of narrow gamma ray radiation lines and the intensity of extremely broadened lines. The goal was to study ways in which high-energy particles are produced in solar flares.
- Hard X-ray spectrometer—to help determine the role that energetic electrons play in the solar flare phenomenon.
- Hard X-ray imaging spectrometer—to image the Sun in hard X-rays and to provide information about the position, extension and spectrum of the hard X-ray bursts in flares.
- Soft X-ray polychromator—to investigate solar activity that produces solar plasma temperatures in the to 0 million degree range and study solar plasma density and temperature.
- Ultraviolet spectrometer and polarimeter—to study the ultraviolet radiation from the solar atmosphere, particularly from active regions, flares, prominences and the active corona, and to study the quiet Sun.
- High altitude observatory coronagraph/polarimeter—to return imagery of the Sun's corona in parts of the visible spectrum as part of an investigation of coronal disturbances created by solar flares.
- Solar constant monitoring package—to monitor the output of the Sun over most of the spectrum and over the entire solar surface.

The satellite was the first solar satellite designed to study a specific solar phenomenon such as flares, using a coordinated set of instruments that measure many different flares in different wavelengths of light.

Test Results—The SMM was launched from the Eastern Test Range (ETR) on 14 February 1980. Subsequently the six science investigations on SMM performed a detailed study of a specific set of solar phenomena: the impulsive, energetic events known as solar flares, and the active regions that are the sites of flares, sunspots and other manifestations of solar activity. One instrument also measured the total output of radiation from the Sun.

An artist conception of 'Solar Max' (*left*) as it proceeded on its mission to study such phenomena as the major short wavelength and coronal manifestations of solar flares. (*Above*) Astronauts George Nelson and James Van Hoften repairing 'Max' in a space shuttle payload bay in April 1984.

They significantly contributed to understanding of the chemistry of the earth's atmosphere, the earth's magnetic field and how solar particles penetrate and become trapped in the magnetosphere.

They provided the first evidence of a region of low-energy electrons enveloping the high energy Van Allen Radiation Region, the first observation of daylight auroras, the first global map of airflow distribution and much other knowledge about magnetic fields, particle radiation, the earth's ionosphere, the shockwave between the geomagnetic field and solar wind that was discovered by Explorer 18, and the hydrogen cloud enveloping the earth.

Beyond the earth, they completed the first sky survey of hydrogen, discovered neutral hydrogen around the Sun, found several strong sources of hydrogen in the Milky Way, and in April 1970, detected a cloud of hydrogen 10 times the size of the Sun around comet Bennett. The existence of large amounts of hydrogen around comets was first discovered around comet Tago-Sato-Kosaka in January 1970 by OAO 2 *Stargazer.* The large amounts of hydrogen around comets that were observed by OAO and OGO clearly establish that hydrogen is a major constituent of comets.

The launch dates of the six OGOs are as follows:

OGO 1	4 September 1961	OGO 4	28 July 1967
OGO 2	14 October 1965	OGO 5	4 March 1968
OGO 3	7 June 1966	OGO 6	5 June 1965

HELIOS SPACECRAFT

Helios 1 and 2, a cooperative project of NASA and the Federal Republic of Germany, were designed to survive and function at distances closer to the Sun than any other spacecraft. Named for the sun god of ancient Greece, the craft were launched on 10 December 1974 and 15 January 1976.

From perihelions lower than 28 million miles on opposite sides of the Sun, they added to knowledge about the solar corona, magnetic field, wind and radiation and about micrometeoroids and other phenomena in this vicinity of space.

SOUNDING ROCKETS, AIRCRAFT, BALLOONS AND GROUND OBSERVATORIES

NASA sounding rockets, high-flying airborne observatories, ground observatories and balloons have contributed volumes of data to NASA scientific programs in the past 25 years. NASA has launched thousands of sounding rockets, many in cooperation with other countries, from ranges in the United States and abroad. Among the contributions of sounding rockets is their ability to explore areas of space too high for balloons and too low for satellites. NASA's fleet of sounding rockets includes Aerobees, Arcas, Argo D-4 (Javelin), Aries, Astrobees, Black Brants, Nike-Apache, Nike-Cajun, Nike-Hawk, Nike-Malemute, Super-Loki and Terrier-Malemute. The most familiar experiments are those involving the release of a chemical that colors the skies.

Observations of the chemical drift provide information on upper-atmospheric winds or magnetic field structure. Sounding rockets are also used in astronomical studies and to test instruments for future use in satellites and other spacecraft.

NASA balloon experiments are designed to study the atmosphere and to make astronomical studies. They are frequently conducted in association with other countries. Among recent balloon studies are those in conjunction with the Solar Mesosphere Explorer to investigate depletion of the ozone layer in our atmosphere. NASA launches its balloons from the National Balloon Facility in Palestine, Texas.

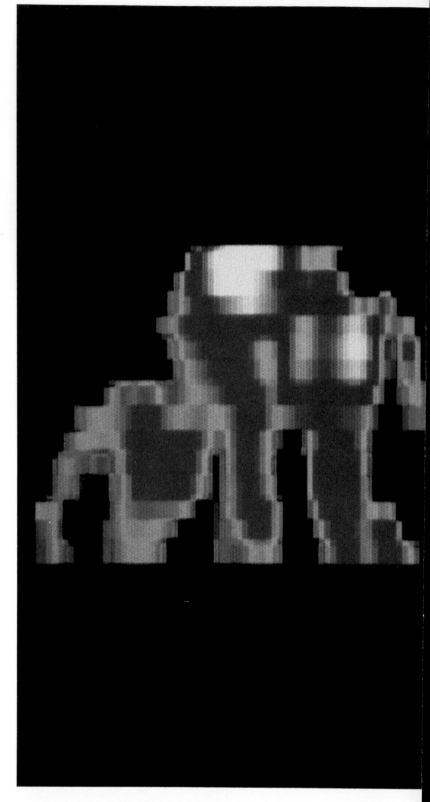

NASA airborne observatories have participated in observations of solar eclipses and other astronomical phenomena, atmospheric studies and geologic and oceanographic observations. Among their most dramatic discoveries were the unexpectedly high ratio of oxygen in the earth's upper atmosphere by the Galileo airborne observatory in 1969 and the five faint rings around Uranus by the Kuiper airborne observatory in 1977.

NASA also supports ground observatories. In 1976 ground observatories discovered a satellite circling Pluto and ascertained that methane ice covered the planet's surface. With these observations, astronomers were able to increase their accuracy in measuring Pluto's mass. Their calculations resulted in a substantial reduction of estimates of Pluto's mass and indirectly increased support for the theory that a massive solar system object existed

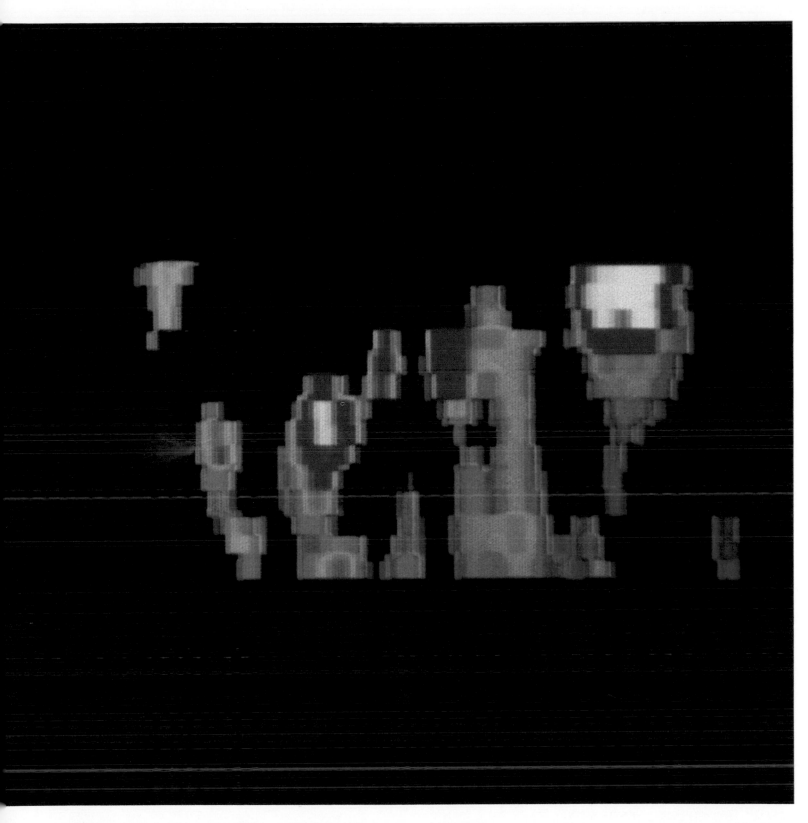

Cloud images from the Infrared Astronomical Satellite (IRAS) showing Magellanic Clouds (*above*), a galaxy near the Milky Way and Cirrus Clouds (*overleaf*) above our own Earth.

beyond Pluto. The reason is that Pluto's mass, as currently estimated, is far less than adequate to produce the perturbations in the orbits of Uranus and Neptune that inadvertently led to the search for and discovery of Pluto.

In 1982 ground observatories reported that Pluto has an atmosphere of gaseous methane. The atmosphere may be the result of Pluto being closer to the Sun as its elliptical orbit carries it inside of the orbit of Neptune.

In 1983 ground observatories discovered a quasar 12 billion light years away, the most distant object discovered. The search that resulted in this discovery started 10 years ago using antennas of NASA's deep-space tracking and data-acquisition network and observatories in the United Kingdom and Australia.

NASA's Project SETI, Search for Extraterrestrial Intelligence, was begun in 1960. It not only searches for radio signals that could be from intelligent creatures, but also does radio astronomy mapping of the sky and studies man-made radio interference that could affect space tracking, data acquisition and communication.

Airborne and ground observatories, sounding rockets and balloons are frequently used jointly, such as in solar eclipse observations. Scientists want to study the solar corona normally masked by bright sunlight. They want to study changes in the atmosphere, magnetic field and other phenomena over all areas of the earth on the ground track of the solar eclipse caused by the sudden cutoff and re-emergence of sunlight.

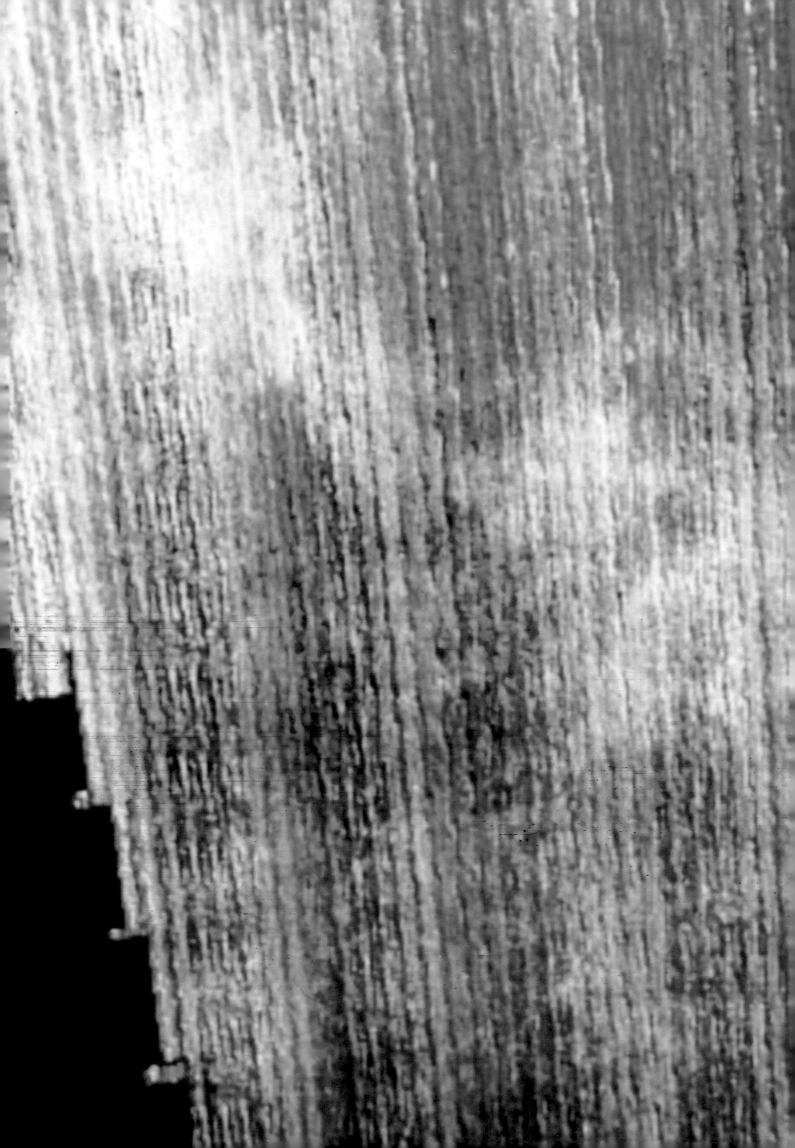

Pioneers In Space

S un-orbiting Pioneers have contributed volumes of data about the solar wind, solar magnetic field, cosmic radiation, micrometeoroids and other phenomena of interplanetary space. Pioneer 4, launched 3 March 1959, was the first United States spacecraft to go into solar orbit, and it yielded excellent radiation data. Pioneer 5, launched 11 March 1960, confirmed the existence of interplanetary magnetic fields and helped explain how solar flares trigger magnetic storms and the northern and southern lights (auroras) on Earth. The satellite also showed that the Forbush decrease of intergalactic cosmic rays near Earth after a solar flare was the same in interplanetary space, and thus does not depend on an Earth-related phenomenon such as the geomagnetic field.

Pioneers 6 through 9, launched 16 December 1965, 17 August 1966, 13 December 1967 and 8 November 1968, supplied volumes of data on the solar wind, magnetic and electrical fields, and cosmic rays in interplanetary space. Pioneer 7 detected effects of the earth's magnetic field more than 3 million miles outward from the night side of the earth. Pioneers 6 to 9 data also drew a new picture of the Sun as the dominant phenomenon of interplanetary space. They found that the solar wind continues well beyond the orbit of Mars. (Pioneer 10, the first Jupiter explorer, continued to report its existence as it crossed the orbit of Pluto.) Analyses of their data indicated that the solar wind is an extension of the

solar corona, the Sun's atmosphere. They revealed that the wind draws out the Sun's magnetic field to form what were previously called interplanetary magnetic fields but now is referred to as the heliosphere. They showed that the combination of the solar wind's outward pull and the rotation of the Sun caused the lines of forces of the magnetic field to be twisted like streams of water from a whirling lawn sprinkler.

Pioneer data showed that solar cosmic rays spiral around the lines of force of the Sun's magnetic field. This indicates they travel through space in well-defined streams.

Planetary spacecraft that made close-range observations of other planets during flybys from orbit and from the surface have significantly enhanced knowledge about the other planets in the solar

The Pioneer 10 spacecraft being raised into a space simulation chamber which subjected the craft to the heat, cold, and simulated radiation it would soon encounter in space. Launched in March of 1972 Pioneer 10's primary objective was to take the first close-up look of Jupiter and return data on about twenty aspects of the planet, its moons and environment. In addition, Pioneer 10 also explored a curving strip of space 620 million miles long extending from the Earth's orbit to Jupiter. Pioneer 10 carried a plaque containing symbols that are expected to be decipherable by intelligent life should it happen to encounter Pioneer. In 1987 Pioneer will be the first man-made object to leave our solar system.

system. They have demonstrated the uniqueness of our own planet. They have provided among other things excellent studies of how, after a presumably common origin, celestial bodies react to different environmental conditions. They suggest the earth's past and future and possibly how to improve the earth's environment.

MARINER SPACECRAFT

Mariner 2 was America's first successful planetary spacecraft. Launched 27 August 1962, it flew past and made close-up observations of Venus on 14 December 1962. Its data supported earth-based microwave scans that suggested a surface temperature as high as 800° F, hot enough to melt lead on both day and night sides. It detected no openings in the dense clouds enveloping Venus. Its data indicated no intrinsic Venusian magnetic field nor increase in radiation. This suggested that Venus has no radiation belt like the Van Allen Radiation Region around the earth. The data are consistent because the Van Allen Radiation Region is attributed to the capture of energetic particles by the earth's magnetic field. Mariner 2 also confirmed the predominance of the solar wind as a feature of interplanetary space and the ubiquity of interplanetary magnetic fields, which scientists now realize are an extension of the Sun's magnetic field dragged out into space by the solar wind.

Data about the environment of interplanetary space have been provided by interplanetary spacecraft and by planetary spacecraft before and frequently after completion of their primary missions.

Ranger 1 (*left*) was the predecessor of the Mariner 1 and 2 spacecraft. The Ranger program was the first of three lunar exploration projects leading to the Apollo program that placed men on the moon. Ranger craft were designed to fly to the moon and return data both enroute and up to the surface of the moon.

Mariner 2, which confirmed that, beneath its bright cloud cover, Venus was a dry, lifeless inferno, also confirmed the existence of the solar wind as a predominant feature of interplanetary space. Pioneer 10, which provided the first close-up of Jupiter, returned data indicating that the Gegenschein—a faint glare in the earth's sky directly opposite the Sun—and the zodiacal light are due to the sunlight reflecting from small particles in interplanetary space rather than in the earth's atmosphere. The zodiacal light is a faint cone of light extending upward from the horizon in the direction of the zodiac, or ecliptic. Pioneer 10 also showed that the heliosphere extends beyond Jupiter.

This section covers only those craft launched to circle the Sun but not to visit a planet. Scientists correlate data from these craft at various points in space with data from planetary spacecraft and scientific earth satellites to increase understanding of the Sun and solar system.

Speeding by Mars on 14 July 1965, Mariner 4, launched 28 November 1964, gave the world its first close look at that planet's surface. The pictures were surprising: a heavily cratered moon-like surface that looked like it may not have changed much in billions of years. Because the pictures covered about 1 percent of Mars, they permitted no conclusions until other spacecraft viewed additional areas of the planet. The pictures covered some areas

continued on page 71

Mariner 1 and 2

Spacecraft Description—Similar to the Ranger series design with a tubular superstructure mounted on a hexagonal base. Two solar panels extended from the base and a high-gain antenna was hinge-mounted below the base. Mariners 1 and 2 weighed 447 pounds and in launch configuration were 5 feet in diameter at the base and 9 feet, 11 inches high. In cruise position with solar panels and high-gain antenna extended, the spacecraft were 16.5 feet across the panels and almost 12 feet high.

Project Objectives—Inject the spacecraft into Venus flyby trajectory; make infrared and microwave measurements of the planet; make fields and particles measurements in interplanetary space and at the planet; communicate experiment data to earth over distances up to 36 million miles.

Spacecraft Payload—The spacecrafts' attitude control system used 10 cold nitrogen gas jets linked with three gyros, an earth sensor and six sun sensors. A hydrazine-fueled midcourse correction motor (50-pound-thrust for up to 43 seconds) was carried to adjust trajectory to an optimum Venus miss distance of 10,000 miles. The communication system consisted of a receiver/transmitter and omnidirectional, high-gain and command antennas; transmitting power was 3 watts. A combination of insulating shields, louvers, paint patterns, aluminum sheet and gold plating was utilized in the thermal control system. 9800 solar cells designed to produce a minimum of 148 watts served as the primary power supply; initial and backup power was provided by a silver/zinc battery. A digital central command computer and sequencer performed all system computation and issued commands in three sequences—launch, midcourse and encounter.

Mariners 1 and 2 carried two planetary experiments: 1) microwave radiometer to determine surface temperature and atmospheric details, 2) infrared radiometer to determine structure of the planet's cloud cover. Interplanetary experiments included: 1) fluxgate magnetometer to measure strength and direction of magnetic fields, 2) ion chamber and three geiger-Mueller tubes to measure intensity and number of energetic particles, 3) cosmic dust detector to measure the flux and momentum of cosmic dust particles, 4) solar plasma detector to measure flow and density of solar plasma and the energy of its particles.

Test Results—Mariner 1 was launched from Kennedy Space Center on 22 July 1962. The vehicle was destroyed by the range safety officer about 290 seconds after launch when it veered off course. The failure apparently was caused by a combination of two factors: improper operaton of the Atlas airborne beacon equipment resulting in a loss of the rate signal from the vehicle for a prolonged period and incorrect data-editing equations.

Mariner 2 was launched into parking orbit from Kennedy Space Center on 27 August 1962, then ejected from the Agena into a Venus flyby trajectory. The midcourse maneuver was successfully accomplished on 4 September, reducing the Venus miss distance from 233,000 to about 20,900 miles. The spacecraft suffered a sudden voltage drop on 30 October and the four interplanetary experiments were turned off. Just as abruptly, the power level returned to normal nine days later and the experiments were turned on again. On 25 November Mariner 2 established a new long-distance communication record, breaking Pioneer 5's deep-space record of 22.5 million miles set in June 1960. Mariner 2 was to pass by the sunward side of Venus on 14 December 1962. The spacecraft's interplanetary experiments provided significant data on a continuous solar wind, reduced cosmic dust density, and particle and magnetic field variations with solar disturbances.

Mariner 4 was launched in November of 1964 on a 228-day mission to Mars. The spacecraft carried instruments for eight interplanetary and planetary experiments including a TV set to record images of the planet. Mariner 4 passed Mars at an altitude of 6118 miles, recording and transmitting man's first close-up pictures of that planet. Planetary science data, including pictures, were transmitted over distances ranging from 134 million to 150 million miles.

Mariner 3 and 4

Spacecraft Description—Octagonal magnesium main body, 50 inches across. High-gain dish mounted atop main structure along with low-gain antenna on top of aluminum tube. Four solar panels extended from top side of octagon, with solar pressure vanes, each 7 square feet of aluminized mylar sheet, attached to end of each solar panel. Spacecraft was 9.5 feet high and 22.6 feet across. Weight: 575 pounds.

Project Objectives—Perform two planetary experiments during Mars flyby (aiming zone centered 8600 miles from surface) and six interplanetary experiments along spacecraft trajectory. Flight also intended to provide engineering experience in spacecraft operation during long-duration flight away from Sun. Mission designed for a flight distance of 350 million miles to reach Mars (about 8½ months flight time) and communication distance of 150 million miles at planetary encounter.

Spacecraft Payload—Single camera to take up to 21 still photos during flyby; playback for transmission in digital form to earth expected to take 8.3 hours per photo. Second planetary experiment, occultation studies of Martian atmospheric pressure, was to be based on spacecraft transmissions as it passed behind planet. Interplanetary experiments included solar plasma probe, ionization chamber and geiger-Mueller tube, solid-state radiation detector and three geiger-Mueller tubes, helium vector magnetometer, cosmic ray telescope and two cosmic dust detectors. Ten-watt transmitter intended to return all data except photos in real time. Hydrazine-fueled engine with 51 pounds of thrust carried for up to two midcourse corrections. Central computer and sequencer provided to control mission sequences.

Project Results—Mariner 3 was successfully placed in parking orbit following launch from Kennedy Space Center on 5 November 1964. Earth escape trajectory was achieved by second Agena burn but spacecraft did not reach planned speed of 25,661 mph and subsequently failed to extend its solar panels and to acquire the Sun and Canopus. Project officials concluded that Mariner 3's fiberglass shroud did not completely jettison as scheduled 5½ minutes after launch. Tests indicated shroud's inner layer separated under combined stresses of aerodynamic heating and rapid pressure drop at altitude. Transmissions ceased nine hours after launch; spacecraft was in solar orbit.

Mariner 4 was successfully launched from Kennedy Space Center on 28 November 1964 and injected into a Mars trajectory with a miss distance of 151,000 miles. Spacecraft acquired Canopus on 30 November, after first locking on several wrong stars. Midcourse maneuver successfully carried out 5 December, with engine fired for 20.06 seconds and trajectory altered so that Mariner 4 will pass behind Mars and within 5400 miles of planet. A second correction was not required. Solar plasma experiment malfunctioned 7 December. The same day, Canopus lock was again lost but was reacquired 17 December. On 10 January the spacecraft was traveling 8624 mph relative to Earth and 70,099 mph relative to the Sun, and had traveled a total of 74.5 million miles (7.5 million miles relative to Earth). Travel distance to Mars was 325 million miles; the planet was 134 million miles from earth during Mariner 4's flyby.

Mariner 5 was launched in June of 1967 on a mission to the planet Venus which took five months. Planetary experiments aboard the spacecraft included a solar plasma probe, radiation detector, helium magnetometer, ultraviolet photometer and S-band occultation based on telemetry signals, Mariner 5 approached to within 2429 miles of Venus, and discovered a 72 to 87 percent carbon dioxide atmosphere without oxygen.

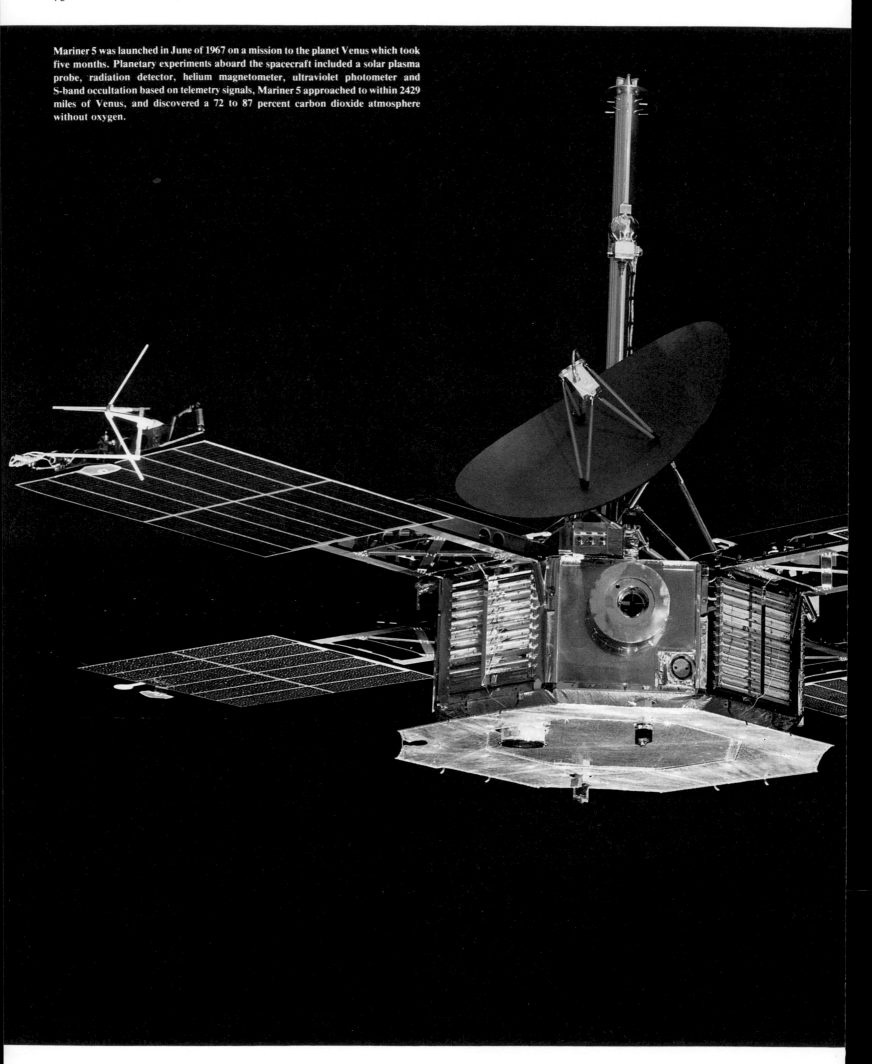

Mariner 5

Spacecraft Description—Modified Mariner 4 backup spacecraft. Main body a magnesium octagon 50 inches across and 20 inches high. Four solar panels deployed from top, high-gain ellipse antenna mounted on superstructure atop body and 88-inch tube on top supported low-gain omnidirectional antenna and helium magnetometer. In flight spacecraft 18 feet across solar panels, 9.5 feet high. Weight: 540 pounds.

Project Objectives—Intended to pass within 2000 miles of Venus to provide data on structure of planet's atmosphere and its radiation and magnetic field environment. Mariner 5 also was designed to return data on interplanetary environment before and after planetary encounter.

Spacecraft Payload—Planetary experiments included S-band occultation based on spacecraft telemetry signals and UV photometry of upper atmosphere atomic hydrogen and oxygen radiation using three photomultiplier tubes. Dual frequency (423.3 and 49.8 mHz) propagation, trapped radiation, magnetic field, solar plasma and celestial mechanics experiments were intended to provide data throughout flight. Data automation system handled data from five experiments.

Project Results—Mariner 5 was launched 14 June 1967 from Kennedy Space Center, placed in parking orbit and then injected into interplanetary trajectory. Sun and Canopus acquisition accomplished without incident. Mid-course maneuver carried out with 17.66-second burn 19 June, reducing Venus miss distance to approximately 2500 miles. Mariner 5 reached Venus on 19 October after 217 million miles of travel. At that time planet was 49.5 million miles from earth. Mariner 5 was to pass Venus ahead of its orbit around Sun, with total occultation time of about 26 minutes. Transmission of science data recorded during flyby was 14 hours after encounter. Mariner 5's post-encounter orbit brought it closer to Sun than any previous probe.

crossed by the supposed Martian canals but showed no readily apparent straight-line features that could be interpreted as artificial. Among other data from Mariner 4 were indications that the surface atmospheric pressure on Mars was less than 10 millibars. Earth's sea-level air pressure is about 1000 millibars. Humans would therefore need a pressure suit on Mars. Mariner 4 also gave additional information about Mars' size, gravity and path around the sun. It detected neither a Martian magnetic field nor radiation belt, but it revealed a Martian ionosphere.

Mariner 5 was launched 14 June 1967 to refine and supplement data about Venus obtained from Mariner 2 and other observations. It contained improved instrumentation, and in October 1967 flew within 2500 miles of Venus as opposed to the 21,645-mile closest approach of Mariner 2. Among the conclusions drawn after this flyby were that Venus' atmosphere is at least 80 percent carbon dioxide and about 100 times denser than the earth's and that Venus' surface temperature may be as high as 800° F. It was also concluded that the solar wind is diverted around Venus by the planet's ionosphere (Earth's magnetic field diverts the solar wind around our planet); the Venusian exosphere, like the earth's, is

continued on page 77

Mariner 6 and 7

Spacecraft Description—(Mariner 6 and 7 spacecraft identical.) Base structure was a 37-pound octagonal forged-magnesium frame, 54.5 inches diagonally and 18 inches deep, containing eight subsystem compartments that also provided structural strengthening. Four solar panels 84 inches long and 35.5 inches wide attached to the top; each panel had 20.7 square feet of solar cells (83 square feet total, each spacecraft); when deployed, they spanned 19 feet. Attitude control jets mounted at panel tips. Low-gain omnidirectional antenna mounted atop four-inch-diameter aluminum tube, which served as a waveguide and extended 88 inches from the top of the base. Cone-shaped thermal control flux monitor also mounted at top of mast. Total height, 11 feet from top of low-gain antennas mast to bottom of lower experiment-mount scan platform. Weight: 910 pounds.

Project Objectives—Primary mission objectives for both Mariner 6 and 7 Mars 1969 flybys were the scientific study of the surface and search of extraterrestrial life, and to develop technologies for future Mars missions. Flights also to further demonstrate engineering concepts and techniques required for long-duration flight away from the sun. Flight duration and distance for Mariner 6 was 156 days and 226 million miles; for Mariner 7, 133 days and 193 million miles; communication distance to Earth during planetary encounters of each spacecraft, respectively, 59.5 million miles (about 5.5 light-minutes) and 61.8 million miles. Both Mariners flew by Mars at about 2000 miles—Mariner 6 passed over the Martian equator on 31 July 1969, concentrating on a belt roughly 90° East of the 1965 photo path of Mariner 6. Mariner 7 overflew the southern hemisphere near the south polar cap on 5 August 1969, after overlapping the Mariner 6 path over the permanently dark area near the equator. Whereas Mariner 4's photos covered 600,000 square miles (1 percent) of the planet's surface, Mariner 6 and 7 far-encounter photos covered all of the planet as it rotated twice for each set of cameras on both vehicles; also, close-ups were made of about 20 percent of the planet's surface. Best-resolution of the approach-phase TV pictures was 15 miles, versus 100-mile-limit from Earth; highest resolution in surface pictures about 900 feet, versus two miles for Mariner 4.

Spacecraft Payload—Both Spacecraft identical. Two television cameras: Camera A, for medium-resolution (wide-angle) approach pictures, equipped with red, green and blue filters to delineate corrected-color differences of the planetary atmosphere and surface; Camera B, for high-resolution (narrow-angle) pictures, programmed to overlap specific areas within regions studied by Camera A, equipped with a yellow filter to reduce the effects of planetary atmospheric haze, and with a modified Schmidt Cassegrain telescope to be used for approach pictures. Camera A similar to the camera flown on Mariner 4, but with a wide-angle lens to cover an area 12 to 15 times larger than for Mariner 4 yet with the same 2-mile resolution quality. Camera B's optical resolution 10 times sharper than Camera A and designed to cover an area 100 times larger on the Martian surface. Alternate camera operation, each taking one picture every 42.25 seconds within the experiment range of 6000 to 2000 miles. Cameras capable of focus adjustments to account for surface altitude variations up to 8.3 miles.

Other experiment packages, which together with cameras were mounted on motor-driven, two-degree-of-freedom scan platform below the octagonal base, included the following: Infrared radiometer (IRR) to perform thermal mapping, overlapping the areas covered by TV experiments; two detectors in each instrument to provide 30 readings, one every 63 seconds; one detector to operate in the range near 300°K, the other around 140°K. IRR's on both spacecraft to scan the Martian surface from late morning to late evening, yielding cooling rates and indices of surface compositions, especially valuable from the dark side not seen by Earth. Chemical constituents of upper atmosphere (60 to 600 miles) to be identified and measured, as well as backup density and temperature data to be provided by ultraviolet spectrometer, the first attempt to use this technique to identify Martian atmospheric gases. Lower atmosphere and surface compositions to be measured by an infrared spectrometer (IRS), celestial mechanics experiment to be performed also, using radio signals during flight, at encounter, and the flight behind and beyond Mars; objectives included determining the mass of Mars, Earth-Mars distance at encounter, and the Earth-Moon mass ratio. Occultation experiment utilized S-band radio signals from the spacecraft, to obtain precise measurements of the radius of Mars, the reflection of radio signals from the planet's surface, and the electron density of its atmosphere.

Primary contents of the eight compartments in base of each spacecraft were: (1) power conversion equipment; (2) midcourse correction propulsion system; (3) central computer and sequencer and attitude control subsystem; (4) telemetry and command subsystem; (5) tape recorders; (6) radio receiver and transmitter; (7) science instrument electronics and data automation subsystem; (8) power booster regulators and nickel cadmium battery.

Project Results—Mariner 6: Launched 25 February 1969 from Kennedy Space Center in a direct-ascent single-burn ballistic trajectory with an initial heliocentric injection velocity of 25,700 mph. About 30 minutes after launch, the spacecraft was rotated from random attitude, to acquire the sun. About 4 hours out, the vehicle was again rotated to lock on Canopus, and space-stabilized with solar panels perpendicular to the sun for the remainder of the flight. A midcourse trajectory correction was made on 28 February while the spacecraft was 750,000 miles out. Mariner 6 was ahead of Mars at launch but affected by solar gravity, slowed to 17,633 mph relative to Mars so that Mars passed the spacecraft and slightly ahead of it at encounter. Mariner 4 crossed the orbit of Mars within 2120 miles, 31 July 1969, on its equatorial flyby, and subsequently flew behind the planet for about 25 minutes.

There were three encounter experiment phases: far-encounter, near-encounter, and occultation and playback. Far-encounter began at encounter minus 54 hours (E-54) and ran to E-7 hours. The far-encounter picture-taking sequence for Mariner 6 ranged from E-48 to E-7 hours and was preceded by a period when the scientific instruments, data systems, and telemetry were turned on. The narrow-angle Camera B took 33 full-disk analog pictures of the

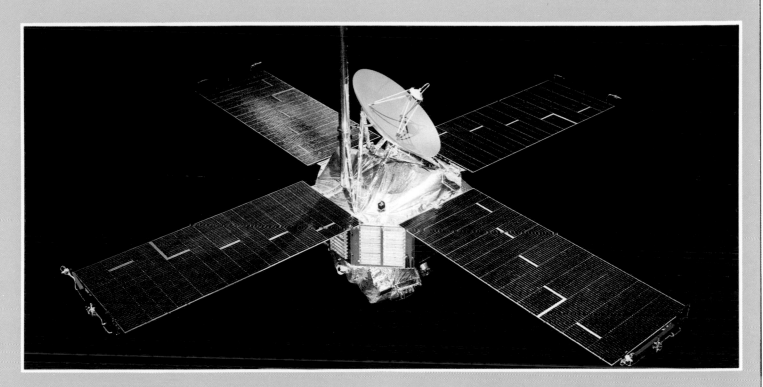

planet, which were then fed to Earth, followed by a series of 17 pictures recorded from E-22 and E-7 hours. During an 18-minute near-encounter period, 22 pictures from both cameras A and B were taken, for a total of 75 for this flyby. The occultation experiment began about E-11 minutes. The absence of a veiling haze, seen earlier in Mariner 4 pictures, corroborated and countered the theory that the planet had a relatively dense and hazy atmosphere extending to 93 miles.

Mariner 7: Launched 27 March 1969 from Kennedy Space Center in a direct-ascent single-burn ballistic trajectory with an initial heliocentric injection velocity of 25,700 mph. About 30 minutes after launch, the spacecraft was rotated from random attitude, to acquire the sun. About 4 hours out, the vehicle was again rotated to lock on Canopus, and space-stabilized. A midcourse trajectory correction was made on 8 April, while Mariner 7 was 2.5 million miles out. Mariner 7 was ahead of Mars at launch but, affected by solar gravity, it slowed to 16,063 mph relative to Mars, so that Mars passed the spacecraft and was slightly ahead of it at encounter. Mariner 7 crossed the orbit of Mars within 2190 miles, 5 August 1969, on its trajectory over the region from the equator to the south polar cap, and subsequently flew behind the planet for over 20 minutes. There were the same three encounter phases for Mariner 7 as for its earlier twin. Far-encounter for Mariner 7 began about encounter minus 72 (E-72) hours, and resulted in 93 far-approach analog picture with Camera B; during near-encounter, 33 pictures were taken with both cameras, for a total of 126 for Mariner 7. Occultation experiment also began at about E-11 minutes. Joint Flight Results: When the Mars flyby mission phase ended, both spacecraft continued to transmit, their 20 watt outputs being used in continuing celestial mechanics experiments. When they passed behind the Sun, tests were made of the theory of electromagnetic radiation bending when in the vicinity of the large solar gravitational fields. Television experiments: transmitted nine times the number of

This type of spacecraft was used for the Mariner 6 and 7 Mars missions. Launched in February and March of 1969 with flybys in July and August of that year, these Mariners carried two TV cameras, IR radiometer and spectrometer, and UV spectrometer.

proposed far-encounter TV pictures, 20 percent more near-encounter pictures; a total of 1100 analog and digital pictures. Pictures gave little indication of any atmospheric 'cloudiness' except thin aerosol haze, near the bright limb and the polar cap, similar to Earth's tropopause; altitude varies horizontally between 12 and 18 miles; it is layerlike and generally about 5 to 10 miles thick. Three types of terrain were noted: crater, featureless (plains), and chaotic, the latter not found on the Moon, and not comparably found on Earth. Craters ranged from 31 to 200 miles (Nix Olympica). South polar cap measured as several feet thick.

IRR experiment: very successful, revealed South polar cap temperature to be between −180°F to −193°F. Surface daytime temperatures in other regions measured from −63°F to +62°F.; night side readings were −63°F to −153°F. IRS experiment: With Mariner 7, some 250 spectral readings were taken, one every 10 seconds, including sweeps over Meridianii Sinus and Hellespontus regions. Confirmed temperature reading from IRR. Detected virtual 90 percent solid and gaseous carbon dioxide saturation of Martian atmosphere at high altitudes, and on surface; revealed thin, extremely slight content waterized atmospheric 'fog' at all latitudes; refuted earlier theory of possible existence of methane or ammonia content in atmosphere near south polar cap; and detected presence of silicate material solids.

UV Spectrometer: also very successful, confirmed presence of CO_2 at 81 miles above surface. Also detected ionized CO_2, carbon monoxide, atomic hydrogen, and very slight traces of molecular oxygen; no nitrogen or nitric oxide were detected. S-band occultation: both spacecraft tests successful. Corroborated possibility of presence at 15, 31, and 37 miles of solid CO_2. Confirmed previous dynamic estimates of the oblateness of Mars.

Mariner 8 and 9

Spacecraft Description—Basic structure was a 40-pound, eight-sided forged magnesium framework (54.5 inches diagonally, 18 inches deep) with eight electronic compartments. Four solar panels, each 84.5 inches long, 35.5 inches wide, were attached by outrigger structures to the top of the octagon. Each panel had a solar cell area of about 20.7 square feet, or a total cell surface of approximately 83 square feet. With solar panels deployed, the spacecraft had a 'wingspan' of 22 feet, 7.5 inches. Height of the spacecraft from bottom of scan platform to top of the low-gain antenna and rocket nozzle was 7.5 feet. Two sets of attitude control jets, consisting of six jets each for three-axis stabilization, were mounted at the tips of the four solar panels. Two spherical propellant tanks for the liquid-fueled rocket engine were mounted side by side atop the octagonal structure. The two-position high-gain antenna was attached to the spacecraft atop the octagon. The Canopus star tracker assembly was located on the upper ring structure of the octagon for a clear field of view between two solar panels. The cruise sun sensor and sun gate were attached to a solar panel outrigger. Mariner 8 and 9's propulsion system—for small trajectory corrections, deceleration into Mars orbit and trim maneuvers—was capable of at least five starts and produced a continuous

A mosaic of Mars (*left*) as imaged from wide-angle pictures taken by Mariner 6. The east-west swath covers an area 450 miles deep and 2500 miles long about 15° south of the Martian equator.

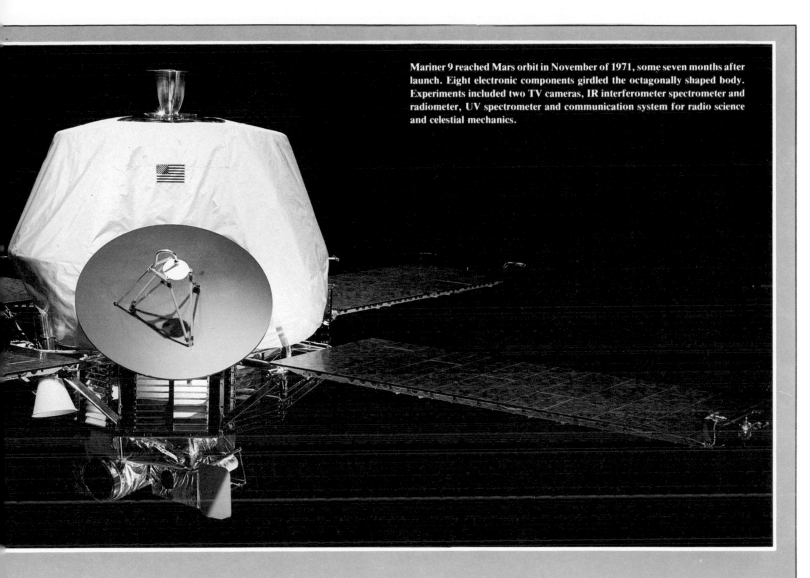

Mariner 9 reached Mars orbit in November of 1971, some seven months after launch. Eight electronic components girdled the octagonally shaped body. Experiments included two TV cameras, IR interferometer spectrometer and radiometer, UV spectrometer and communication system for radio science and celestial mechanics.

thrust of 300 pounds. Science instruments were mounted on a scan platform, which can be rotated about two axes to point the instrument toward Mars during the spacecraft's approach to the planet and while in orbit. Launch weight, including 1000 pounds of fuel and oxidizer, was about 2200 pounds.

Spacecraft Payload—Composed of spacecraft subsystems responsible for gathering, formatting and transmitting data to the earth during periods of scientific observation. These included: TV subsystem, consisting of two cameras and an electronics package, with one camera configured to take wide-angle, low-resolution pictures and the other to take narrow-angle, high-resolution pictures; an Ebert-Fasti type ultraviolet spectrometer to measure incident ultraviolet radiation emitted by gases in the Martian upper atmosphere; an infrared radiometer to measure infrared heat radiation from the Martian surface; an infrared interferometer spectrometer to make a series of interferograms of the planet surface, deep space and warm blackbody sources; a data automation subsystem for sampling, digitizing, formating and output processing of all data from the science experiments; a data storage subsystem to record the high-resolution data stream from the data automation subsystem, store it until an oppor tune phase of the mission, and then play back through the flight telemetry subsystem, which gathers data from many sources and outputs a single composite telemetry signal to the radio system for transmission to Earth.

Project Objectives—After a 167-day trip from Earth and insertion in orbit around Mars, the Mariners' basic 90-day mission was to: radio back to Earth between 25 and 30 billion bits of scientific information about Mars; take about 60 television pictures a day for a total of more than 5000 pictures; take scores of TV pictures of the two moons; map more than 70 percent of the entire Martian surface; study the temperature and composition of the planet's surface with infrared instruments; study the composition and structure of the atmosphere with an ultraviolet instrument, and deter- mine the structure and pressue of the atmosphere by meas- uring changes in Mariner's radio signal as it disappears and reappears behind the planet.

Project Results—Mariner 9 was launched from Kennedy Space Center, 30 May 1971. The spacecraft was inserted into orbit around Mars on 13 November 1971, after a 167-day flight. Upon encounter, a massive dust storm obscured virtu- ally the entire planet. The storm's gradual clearing allowed the spacecraft to begin its scientific mission, and by 6 February 1971, the dust had settled in all areas of the planet except the north polar regions. By that time, Mariner 9 had mapped more than one-third of the planet's surface. Despite the delay, the spacecraft achieved virtually all its initial objec- tives. Project scientists were pleased with the volume of data that was returned from the spacecraft. Mariner 9 combined the original mission objectives of Mariner 8 which failed shortly after launch on 8 May 1971.

An image of the planet Mercury from Mariner 10. This composite photograph clearly shows the cratered surface of the planet and provides planetary scientists with important clues about the age of this planet.

made up largely of hydrogen and Venus has no detectable magnetic field nor radiation belt.

Mariners 6 and 7 were launched on 25 February and 27 March 1969, respectively, and flew as close as 2000 miles to Mars on 31 July and 5 August 1969, respectively. Mariner 6 flew past the planet along its equator. Mariner 7 overlapped part of the Mariner 6 ground track and then sped south over the south polar ice cap. Both spacecraft took pictures of Mars and studied it with infrared and ultraviolet sensors. Their pictures show not only cratered but also smooth and chaotic surfaces. The chaotic region, about the size of Alaska, is characterized by short ridges, slumped valleys, and other irregularities that resemble the aftereffects of a landslide or quake. Nowhere on earth is a comparable feature so vast.

Launched 30 May 1971, Mariner 9 went into Martian orbit on 13 November 1971, the first spacecraft placed into orbit around another planet. It orbited and studied Mars and the planet's two tiny satellites, Deimos and Phobos, until 27 October 1972. It arrived at a discouraging time when a dust storm enveloped most of the planet. Even the dust storm, however, provided information of value such as the atmospheric circulation pattern and the fact that only on Mars were dust storms of such magnitude observed. When the storm cleared, Mariner 9 was able to photograph Martian geography in remarkable detail. Its photographs show Martian volcanic mountains, including one larger than any on Earth; canyons including one that would stretch across the United States from the Atlantic to the Pacific Ocean; and signs that rivers and possibly seas may have existed on Mars.

Mars' two small satellites, Deimos and Phobos, appear as points of light in ground observatory telescopes. Mariner 9 swept close to both, providing pictures that showed them to be irregularly shaped and heavily cratered.

MARINER 10

Mariner 10, launched 3 November 1973, flew by Venus on 5 February 1974 and in a solar orbit, swept nearby and gathered information about Mercury on three separate occasions: 29 March and 21 September 1974 and 16 March 1975. These first close-ups of Mercury reveal an ancient surface bearing the scars of huge meteorites that crashed into it billions of years ago. They show unique large scarps (cliffs) that appear to have been caused by crustal compression when the planet's interior cooled. A Mercurian magnetic field, about a hundredth the magnitude of the earth's, and an atmosphere, about a trillionth of the density of the earth's, were detected. The Mercurian atmosphere is made up of argon, neon and helium. Data suggest a heavy iron-rich core making up about half the planet's volume. (Earth's core is about 25 percent of its volume.) Mariner 10 reported that Mercury's surface temperatures were 950° F on the sunlit side and −350° F on the night side.

Mariner 10 was the first picture-taking spacecraft to view Venus. However, its optical cameras failed to find an opening in the clouds that shroud the planet. Mariner 10's ultraviolet cameras revealed that Venus' topmost clouds circled the planet 60 times faster than Venus rotates. They also confirmed a long-held theory about the earth's weather, that solar heat causes air to rise in the tropical area, flow to the poles, cool, fall and then return to the tropics where the process is repeated. No such process could be discerned on Earth as Earth's rapid rotation, variable atmospheric water content, sizable axial tilt and mixing of continental and ocean air masses produce strong air currents obscuring the equatorial-polar flow. Venus has practically no water vapor, rotates slowly, has no axial tilt, and no mixing of continental and ocean air masses (no oceans) to obscure this flow.

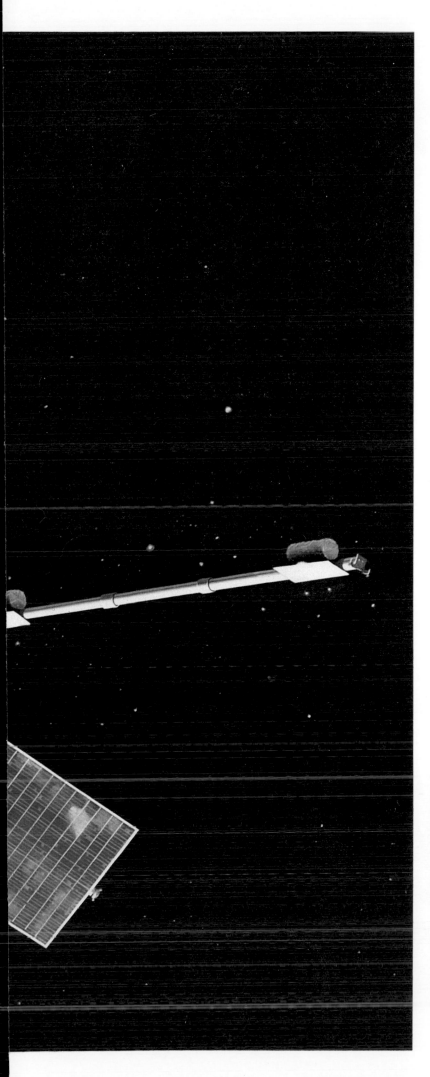

Mariner 10

Spacecraft Description—Mariner 10's basic structure was a 40-pound, eight-sided forged magnesium framework with eight electronics compartments. The framework was 54½ inches diagonally and 18 inches deep. Two solar panels, each 106 inches long and 38 inches wide, were attached by outriggers to the top of the octagon. The panels supported 54.9 square feet of solar cell area. A spherical propellant tank occupied the central cavity of the octagon; the nozzle of the spacecraft's 50-pound-thrust, liquid monopropellant hydrazine engine projected from the sunward side of the octagon through the center of a deployable heat shield. A Canopus star tracker was located on the upper ring of the octagon. Acquisition sun sensors were at the tips of the solar panels. A sunshade, deployed after launch, protected the side of the spacecraft facing the sun. Launch weight was about 1108 pounds, including 64 pounds of fuel and nitrogen gas, and 172 pounds of instruments.

Project Objectives—Mariner 10 was to fly past Venus, taking the first pictures of the planet and measuring its ultraviolet markings. Using Venus' gravity, Mariner 10 was then to fly on to Mercury, taking more pictures and providing man's first close view of the planet nearest the Sun. Venus flyby was 5 February 1974, and the first Mercury flyby was 29 March 1974. A second flyby past Mercury was 21 September 1974. More than 8000 pictures of the two planets were returned.

Spacecraft Payload—Mariner 10 carried instrumentation for six experiments. A seventh—celestial mechanics and radio science—used signals from the spacecraft's transmitters, S-band and X-band. Two identical television cameras and their telescope assemblies were mounted on a motordriven scan platform on the shaded side of the spacecraft. An ultraviolet airglow spectrometer was positioned alongside the TV cameras. A second element of the ultraviolet experiment, the UV occultation spectrometer, and a charged particle telescope were body-fixed to the octagon. A 20-foot-long hinged boom supported two magnetometers, one at the end and one about 13 feet from the spacecraft. The boom was deployed perpendicular to the spacecraft's long axis. Plasma detectors were mounted on a short, motor-driven boom to provide unobstructed viewing beyond the spacecraft end of the boom.

Project Results—Mariner 10 was successfully launched from Kennedy Space Center 3 November 1973. It took test pictures of Earth and the Moon during early hours of the flight. Some problems were encountered with heaters on Mariner 10's cameras, and the cameras were left on to prevent freezing during the interplanetary cruise.

An artist's conception of the Mariner 10 spacecraft voyaging toward the Sun in search of Venus and Mercury, the inner planets of our Solar System. Mariner 10, a basic Mariner design with some modifications for inner solar system flight, flew NASA's first dual planet mission. The eleven hundred pound spacecraft was launched in November of 1973, passed Venus in February of 1974 and encountered Mercury fifty-two days later. The scientific equipment carried onboard included two television cameras, infrared radiometer, extreme ultraviolet spectrometer and an airglow instrument, magnetometer, plasma science experiments, charged particle telescope and radio science experiments. Mariner 10 returned some 8000 photographs as well as other scientific data.

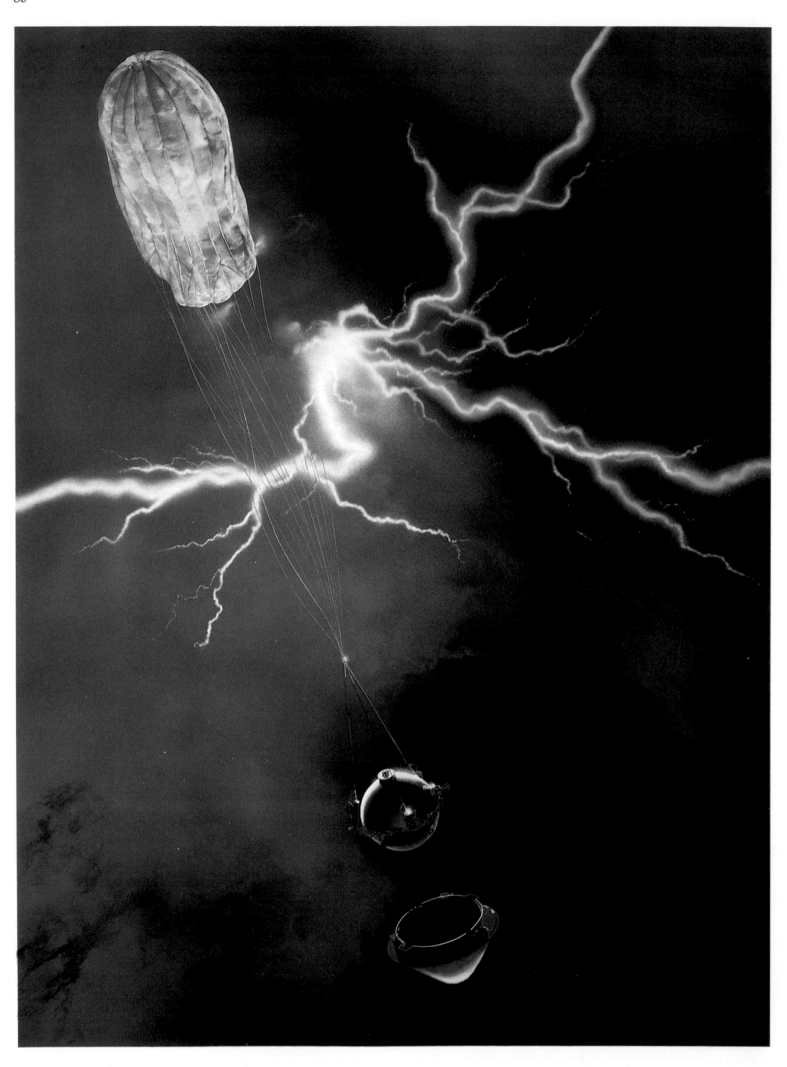

PIONEER-VENUS MISSION

The Pioneer-Venus mission included an orbiter, launched 20 May 1978 and placed into Venusian orbit on 4 December 1978, and a multiprobe bus, launched 8 August 1978, that separated about three weeks before entry into Venus' atmosphere into four probes and the bus. The five entered Venus' atmosphere at widely separated locations on 9 December 1978 and returned information as they descended to the surface. Although none was designed to survive landing, one probe transmitted data for an hour afterward.

Orbiter radar data provided a topographic map of about 90 percent of Venus showing that most of Venus is a gently rolling plain. There are two prominent plateaus: one as large as Australia, the other as large as the upper half of Africa. There is a volcanic structure larger than the earth's Hawaii-Midway chain—Earth's

An artist's conception of a Pioneer Venus probe (*left*) as it deploys in the hostile Venusian atmosphere. The round cylinder is the pressure module with the heatshield aeroshell structure falling away. An infrared image of Venus (*below*) in 1978. This computer enhanced image displays three distinct regions of atmospheric structure. The polar region is characterized by a cloud top temperature of -10°F (250 K) while the temperature of the cloudtops in the mid-latitudes was at -28°F (240 K) with a cool region of enhanced cloudiness surrounding the pole with temperatures of -72°F (215 K). Such studies of atmospheric structure provide scientists with a three-dimensional view of Venus's meteorology.

largest—and a mountain that towers higher over Venus' great plain than Earth's Mount Everest over sea level. Other data indicated two major volcanic areas that vent the planet's internal heat. This makes Venus the third solar system body—the others are Earth and Jupiter's satellite Io—with significant volcanic activity.

Orbiter and probe data refined information about Venus' atmosphere. They showed that the temperature at Venus' surface is 900° F and air pressure on Venus is about 100 times greater than the earth's sea-level pressure. The composition of Venus' lower atmosphere is 96 percent carbon dioxide, 3 percent nitrogen and 1 percent other gases, including extremely small parts of sulphur dioxide and water vapor. Venus' clouds are composed of three distinct layers, all of which consist mostly of corrosive sulfuric acid droplets.

Pioneer-Venus confirmed that the greenhouse effect is responsible for Venus' inferno-like surface temperatures; it also supported information about atmospheric properties and the absence of an intrinsic magnetic field observed by previous spacecraft. It discovered an excess of primordial gases, compared with Mars and Earth, that seem to conflict with theories of planetary evolution. According to these theories, the gases should be less abundant on a planet that formed closer to the Sun than on one that formed farther away.

continued on page 84

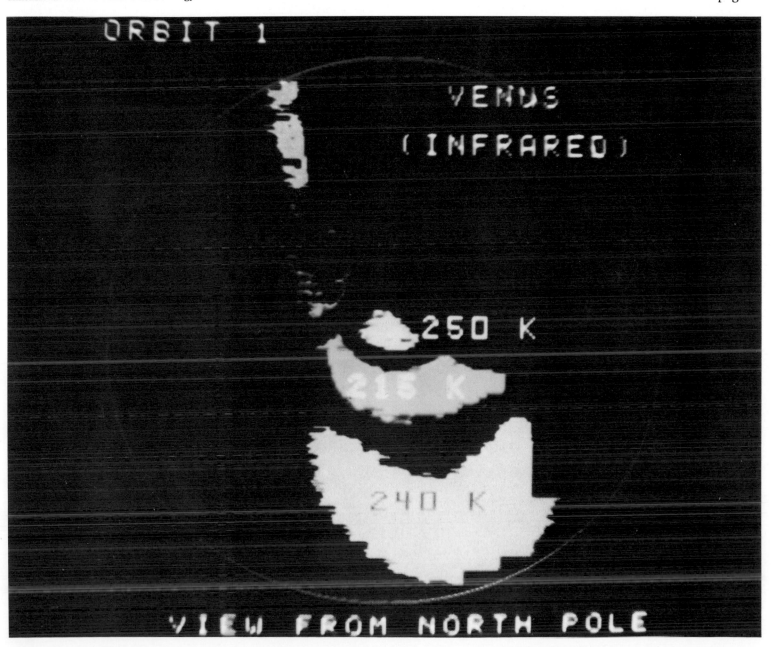

Pioneer Venus Spacecraft
Orbiter

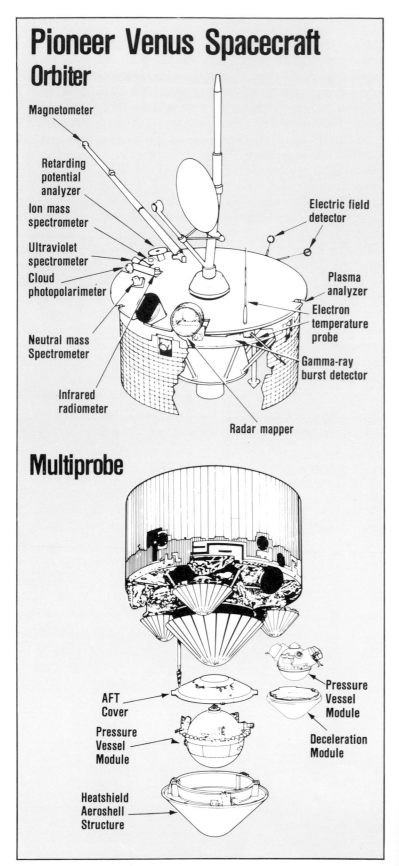

- Magnetometer
- Retarding potential analyzer
- Ion mass spectrometer
- Ultraviolet spectrometer
- Cloud photopolarimeter
- Neutral mass Spectrometer
- Infrared radiometer
- Electric field detector
- Plasma analyzer
- Electron temperature probe
- Gamma-ray burst detector
- Radar mapper

Multiprobe

- AFT Cover
- Pressure Vessel Module
- Heatshield Aeroshell Structure
- Pressure Vessel Module
- Deceleration Module

An artist's conception (*right*) of the hostile environment of Venus. The Pioneer Venus spacecraft (*facing page*) being readied for the trip to the planet Venus. An engineer at Hughes Aircraft Company in El Segundo, California is seen preparing the orbiter (*foreground*) for the May 1978 launch date. The multiprobe spacecraft in the background was readied for an August 1978 launch. The orbiter completed its 300 million mile voyage in just over six months reaching orbit in December of 1978, less than a week before the multiprobe. The orbiter's primary mission was to observe the planet for the length of time it takes Venus to make one revolution on its axis. The multiprobe in the background (made up of four probes) was to enter the atmosphere at various places to measure atmospheric and surface phenomena. The probes were designed as hard landers and were not expected to survive the impact with the planet surface. Luckily, one of the probes did survive and transmitted data for 67 minutes.

Pioneer Venus-A (Orbiter)

Spacecraft Description—Pioneer Venus-A (Orbiter) and Pioneer Venus-B (Multiprobe) spacecraft shared a basic modular design. The basic portions of both spacecraft were their main bodies, flat cylinders, 8.2 feet in dimater and 3.8 feet high. These provided a spin-stabilized platform for scientific instruments and spacecraft systems. A solar array was attached to the equipment shelf. The Venus Orbiter spacecraft incorporated a despun, high-gain dish antenna on a 10-foot mast, and sensor elements mounted on booms. Magnetometer sensors were mounted on a three-section, deployable 15-foot boom. Launch weight of the Orbiter was about 1283 pounds, with 99 pounds of scientific instruments. Weight after orbital insertion was 811 pounds.

Project Objectives—The two Pioneer flights to Venus were to explore the atmosphere of the planet, study its surface using radar, and determine its global shape and density distribution. Pioneer Venus A, the Orbiter, was placed in a 24-hour orbit around Venus for remote-sensing and direct measurements. At orbital low point, Orbiter came within 93 miles of the surface of Venus, sampling the composition of the atmosphere and making radar measurements of the surface elevations. Daily ultraviolet and infrared pictures of the clouds and atmosphere were also taken. The spacecraft also made a variety of other remote-sensing and direct measurements of the planet and its surrounding environment. The satellite's primary mission had been selected to cover the time it takes Venus to make one revolution on its axis—243 Earth days.

Spacecraft Payload—Pioneer Venus-A carried 12 scientific instruments, all of which were mounted directly on the top side of the equipment shelf. They were: a cloud photo polarimeter, a surface radar mapper, an infrared radiometer,

an airglow ultraviolet spectrometer, a neutral mass spectrometer, an ion mass spectrometer, a solar wind plasma analyzer, a magnetometer, an electric field detector, electron temperature probes, a charged particle retarding potential analyzer and a gamma ray burst detector.

A despun, high-gain parabolic-reflector antenna focused a 7.6-degree-wide radio beam on the earth throughout the mission. The Orbiter carried a million-bit memory and a data-handling system.

Project Results—Pioneer Venus-A was launched from Kennedy Space Center on 20 May 1978. It was injected into orbit around Venus on 4 December 1978 after completing its 300 million-mile voyage.

Pioneer Venus-B (Multiprobe)

Spacecraft Description—Pioneer Venus-B, a multiprobe, was made up of a transporter bus, a large probe, and three identical smaller probes. All were to enter the atmosphere of Venus at various points, and all carried instruments and sensors to measure atmospheric and surface phenomena. The transponder was a spin-stabilized, short cylinder 7.6 feet in diameter housing instruments, communications, and navigation systems. All four probes were geometrically similar. The main component of each housed instruments, communications, data, command and power systems. The four probes were launched from the multiprobe 7.8 million miles from the planet to fly to their entry points, two on the day side and two on the night side of Venus. The multiprobe spacecraft weighed 1993 pounds and carried 112 pounds of scientific instruments.

Project Objectives—The multiprobe divided into five atmosphere entry craft as it approached Venus from Earth: the transporter unit (bus), the large probe and three small probes. The probes measured Venusian atmosphere from its uppermost levels down to the superheated regions at the surface. After launching the probes, the bus, too, entered and measured the composition of Venus' upper atmosphere before burning up. The objectives of these probes was to determine the atmospheric structure of Venus from 124 miles to impact at four entry sites well separated from one another. Temperature, pressure and acceleration sensors on all four probes were designed to yield data on the location and intensities of atmospheric turbulence, the variation of temperatures with pressure and altitude, the weight and the radial distance to the center of Venus.

Spacecraft Payload—The large probe contained seven scientific instruments: a neutral mass spectrometer, a gas chromatograph, a solar flux radiometer, an infrared radiometer, a cloud particle size spectrometer, a nephelometer and pressure/temperature/acceleration sensors. The three small probes were identical, 2.5 feet in diameter, weighing 198 pounds. Like the large probe, each of the small probes consisted of a forward heat shield, a pressure vessel and an afterbody. Each carried a nephelometer, a net flux radiometer, and pressure/temperature/acceleration sensors. Each probe also carried a communications system consisting of a solid state transmitter and a hemispherical coverage antenna. All four probes contained a command system, data handling system and power system. The multiprobe bus carried a neutral mass spectrometer and an ion mass spectrometer, which provided the mission's only high upper atmosphere composition measurements, operating as the bus entered the atmosphere but before it started to burn up.

PIONEERS 10 AND 11

Pioneer 10, launched 3 March 1972, was the first spacecraft to cross the Asteroid Belt, the first to make close-range observations of the Jupiter system, and the first to go beyond the outermost planets. In December 1973 it swept nearby Jupiter, finding no solid surface under the thick and deep clouds enveloping the planet. Thus, the world learned that Jupiter is a planet of liquid hydrogen. It explored the huge Jupiter magnetosphere, made close-up pictures of the Great Red Spot and other atmospheric features, and observed and measured at relatively close range Jupiter's large Galilean satellites—Io, Europa, Ganymede and Callisto.

After passing Jupiter, Pioneer 10 continued to map the heliosphere, the giant solar magnetic field drawn out from the Sun by the solar wind. Pioneer 10 found that the heliosphere, like the magnetospheres of Earth and Jupiter, behaves like a cosmic jellyfish, altering its shape in response to rises and falls in solar activity. It also reported that the speed of the solar wind does not decrease with distance from the Sun. On 13 June 1983 Pioneer 10 crossed the orbit of Neptune, which will be the planet farthest out from the sun for the next 17 years. This is because Pluto, although farthest on average, has an extremely elliptical orbit that crosses and goes inside of Neptune's.

Pioneer 10 is searching for the limits, or outer boundary, of the heliosphere. Together with Pioneer 11, it is also searching for a mysterious massive object beyond the known planets. Scientists

A test model of the Pioneer 10 spacecraft (*above*) mounted on a shake table in order to test Pioneer's ability to withstand vibrations during launch. Pioneer 10's primary mission was to take the first close-up look at the planet Jupiter. The spacecraft was also the first designed to travel into the outer solar system and operate effectively there, possibly for as long as seven years and as far from the Sun as 1.5 billion miles. Pioneer returned data on about twenty aspects of the big planet, its moons and environment, from thirteen experiments carried out by eleven onboard scientific instruments. The Pioneer 11 spacecraft (*right*) during a checkout with a mockup of the launch vehicle's third stage at Kennedy Space Center.

hypothesize the existence of the object because of unexplained irregularities in the orbits of Uranus and Neptune. Pluto's mass is insufficient to cause these irregularities. Considering the possibility, however remote, that Pioneer 10 may encounter intelligent extraterrestrials, NASA equipped Pioneer 10 with a plaque. The plaque has diagrams, sketches and binary numbers indicating where, when and by whom Pioneer 10 was launched.

Launched 6 April 1973, Pioneer 11 passed as close to Jupiter as 13,000 miles, compared to the 40,000 miles closest approach of Pioneer 10 on 4 December 1973. It provided additional detailed data and pictures on Jupiter and its satellites, including the first look at Jupiter's north and south poles, which cannot be seen from Earth. This view was possible because Pioneer 11 was guided so that Jovian gravity actually threw the craft out of the plane of the ecliptic in which the planets lie. From above this plane, Pioneer 11 was able to confirm that the heliosphere extends outward in all directions from the Sun and is broken into northern and southern hemispheres by a bobbing sheet of electric current.

Pioneer 11 swept nearby Saturn on 1 September 1979, demonstrating a safe flight path for the more sophisticated Voyagers to follow through the rings. It provided the first close-up observations of Saturn, its rings, satellites, magnetic field, radiation belts and stormy atmosphere. It showed areas smaller than but similar to the Great Red Spot on Jupiter in Saturn's clouds. Pioneer 11 found no solid surface on Saturn, but discovered at least one additional satellite and ring. Its data suggested that Saturn's rings are icy in composition with little or no rock or metal, and that Saturn's three largest satellites—Titan, Rhea and Iapetus—are composed in large part of ice.

Jupiter (*above*) imaged from Pioneer 11 at a distance of more than one million miles. Pioneer 10 (*right*) mounted on a test stand for calibration and flight readiness checkout. The plaque (*far right, top*) carried Pioneers 10 and 11 containing symbols which are expected to be decipherable by intelligent life. The symbols include the hydrogen atom, one of the most common in the universe; a map of the radio energy originating from our sun so that the stellar source of Pioneer can be identified; a map of the solar system showing the location of the earth with Pioneer leaving it; and stylized drawings of male and female human figures superimposed over a stylized drawing of Pioneer in the same scale. Pioneer (*far right, bottom*) being mated to the upper stage of the launch vehicle. The launch housing, or nose cone of Pioneer may be seen in the background.

Pioneer 10

Spacecraft Description—The spacecraft equipment compartment was a 14-inch-deep flat box, top and bottom of which were regular hexagons 28 inches on a side. One side joined to a smaller box also 14 inches deep whose top and bottom are irregular hexagons. The smaller box contained 37 pounds of on-board experiments. Attached to the hexagonal front face of the equipment compartment was a 9-foot-diameter, 18-inch-deep antenna. High-gain antenna feed and medium-gain antenna horn were mounted at the focal point of the antenna dish on three struts projecting about four feet forward of the rim of the dish. At the rim of the antenna dish was a sun sensor. A star sensor looked through an opening in the equipment compartment and was protected from sunlight by a hood. Both compartments had aluminum frames with bottoms and side walls of aluminum honeycomb sandwich. Rigid external tubular trusswork supported the dish antenna, three pairs of thrusters located near the rim of the dish, boom mounts, and launch vehicle attachment ring. Total weight of Pioneer 10 at launch was 570 pounds, including 65 pounds of scientific instruments.

Project Objectives—To investigate Jupiter and its environment, the asteroid belt, the interplanetary medium perhaps as far out as Saturn's orbit, 2 billion miles from the Sun, then to escape the solar system, the first man-made object to do so.

Spacecraft Payload—Eleven instruments and the spacecraft radio, which was used to conduct two experiments (celestial mechanics and S-band occultation). The instruments included a meteoroid detector, an asteroid/meteoroid detector, a plasma analyzer, a helium vector magnetometer, a charged-particle detector, a cosmic-ray telescope, a geiger-tube telescope, a trapped-radiation telescope, an ultraviolet photometer, an infrared radiometer and an imaging photopolarimeter. The plasma analyzer, cosmic-ray telescope, asteroid/meteoroid telescopes, meteoroid sensors and the magnetometer sensors are mounted outside the instrument compartment.

Project Results—Pioneer 10 was successfully launched 3 March 1972 from Kennedy Space Center. It crossed the Moon's orbit 11 hours after launch. It entered the asteroid belt 15 July 1972 and emerged unscathed 15 February 1973, after a seven-month transit. Planet encounter was on 3 December 1973, and flyby duration lasted for one week.

Pioneer 11

Spacecraft Description—The spacecraft equipment compartment was a 14-inch-deep flat box, top and bottom of which were regular hexagons 28 inches on a side. One side joined to a smaller box also 14 inches deep whose top and bottom were irregular hexagons. The smaller box contained 37 pounds of onboard experiments. Attached to the hexagonal front face of the equipment compartment was a 9-foot-diameter, 18-inch-deep antenna. High-gain antenna feed and medium-gain antenna horn were mounted at the focal point of the antenna dish on three struts projecting about four feet forward of the rim of the dish. Two three-rod trusses, 120 degrees apart, projected from two sides of the equipment compartment to deploy the radioisotope ther-

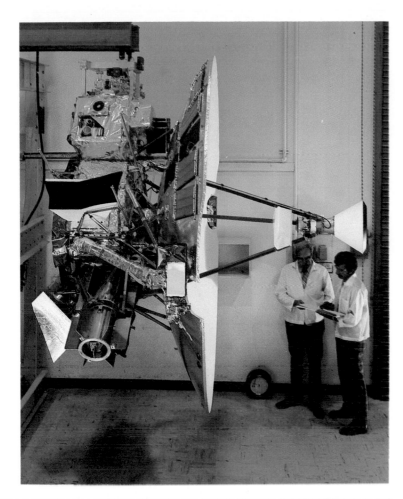

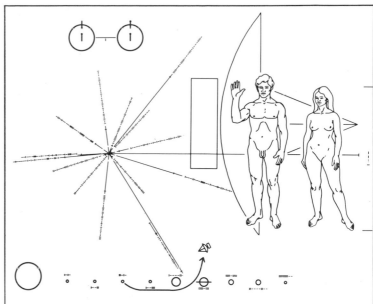

moelectric generator (RTG) power sources about 10 feet from the center of the spacecraft. A third boom, 120 degrees from each of the other two, projected from the experiment compartment and positions the magnetometer sensors 21.5 feet from the spacecraft center. At the rim of the antenna dish was a sun sensor. A star sensor looked through an opening in the equipment compartment and was protected from sunlight by a hood. Both compartments had aluminum frames with bottoms and side walls of aluminum honeycomb. The dish antenna was made of aluminum honeycomb sandwich. Total weight of Pioneer 11 at launch was 570 pounds, including 66 pounds of scientific instruments.

Project Objectives—To investigate the interplanetary medium beyond the orbit of Mars, the asteroid belt, and the near-Jupiter environment.

Spacecraft Payload—Twelve instruments and the spacecraft radio, which was used to conduct two experiments (celestial mechanics and S-band occultation). The instruments included a meteoroid detector, an asteroid/meteoroid detector, a plasma analyzer, a vector helium magnetometer, a charged-particle detector, a cosmic-ray telescope, a geiger-tube telescope, a trapped-radiation detector, ultraviolet photometer, infrared radiometer, an imaging photopolarimeter and a flux-gate magnetometer.

Project Results—Pioneer 11 was successfully launched 6 April 1973 from Kennedy Space Center. It crossed the Moon's orbit 11 hours after launch. It entered the asteroid belt 18 August 1973 and emerged 18 March 1974. Jupiter encounter was 3 December 1974.

VIKING

Viking made the world's most extensive study of Mars. The project used two spacecraft, each of which had an orbiter and lander.

Viking 1, launched 20 August 1975, went into Martian orbit on 19 July 1976 and put its lander on the surface 20 July 1976. The orbiter stopped transmitting on 7 August 1980, and the lander late in November 1982.

Viking 2, launched 9 September 1975, went into Martian orbit on 7 August 1976. Its lander touched down on 3 September 1976. The orbiter reported until 24 July 1978, and the lander until 12 April 1980. They returned a wealth of photographs and other data, mapping about 97 percent of Mars. Their success generated so much public enthusiasm that about $50,000 was raised in 1980 to prolong the project. Late in 1980, the Viking 1 lander was renamed Mutch Memorial Station in memory of Dr Thomas A Mutch, former Viking Lander Imaging Team leader and former NASA Associate Administrator for Space Science, who disappeared while mountain climbing in the Himalayas.

The Viking spacecraft made numerous significant discoveries about Mars. The Martian atmosphere, although too thin for most living things on Earth to survive (about a 1/100 as dense as

continued on page 117

Lifting upward enroute to Mars, Viking 2 was launched aboard a Titan Centaur in September of 1975, just twenty days after its sistership Viking 1. On a trajectory which brought it into the vicinity of Mars eleven months later, Viking 2 was headed for a landing area in the Cydonia region, a lowlands area near the southern fringe of the north polar cap. Views from Viking on the way to Mars (*overleaf*) and (*on pages 92 and 93*) the Valles Marineris as seen from Viking 1. The objective of the Viking missions was to study Mars from orbit, and to land two landers to study the surface composition and to search for life on the planet.

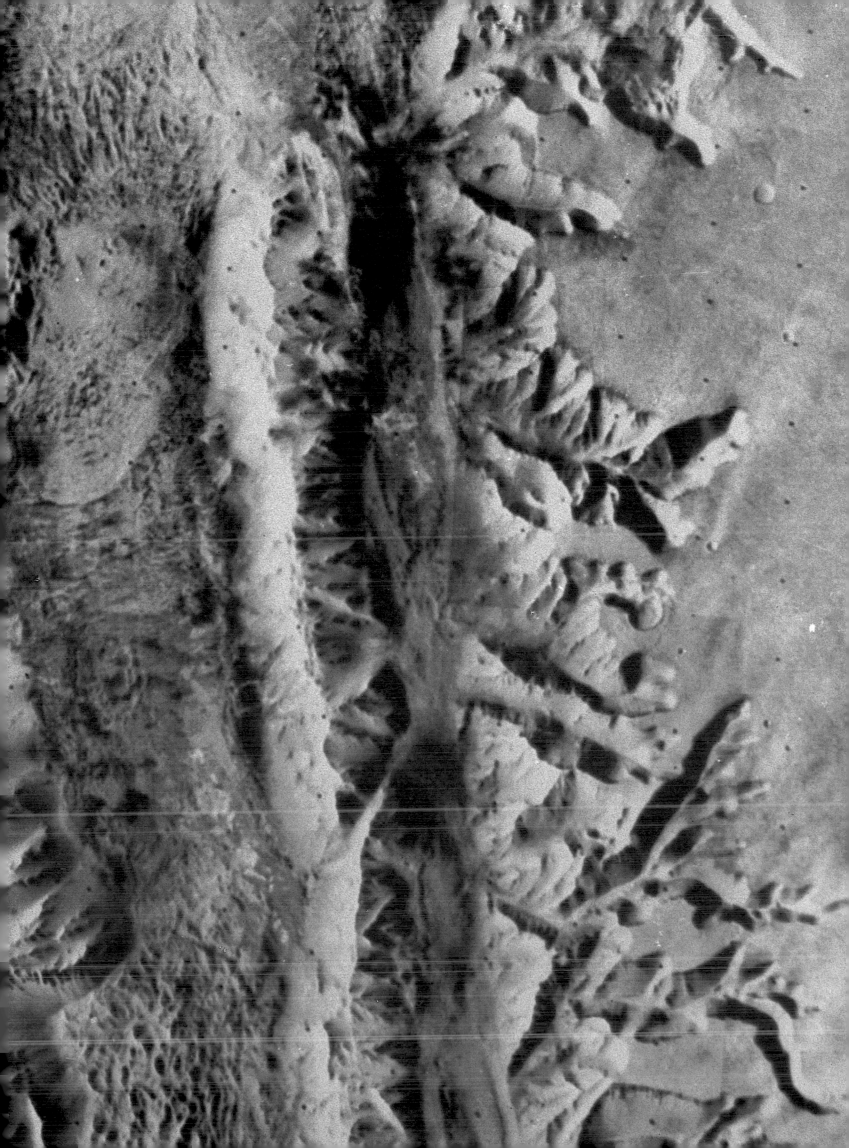

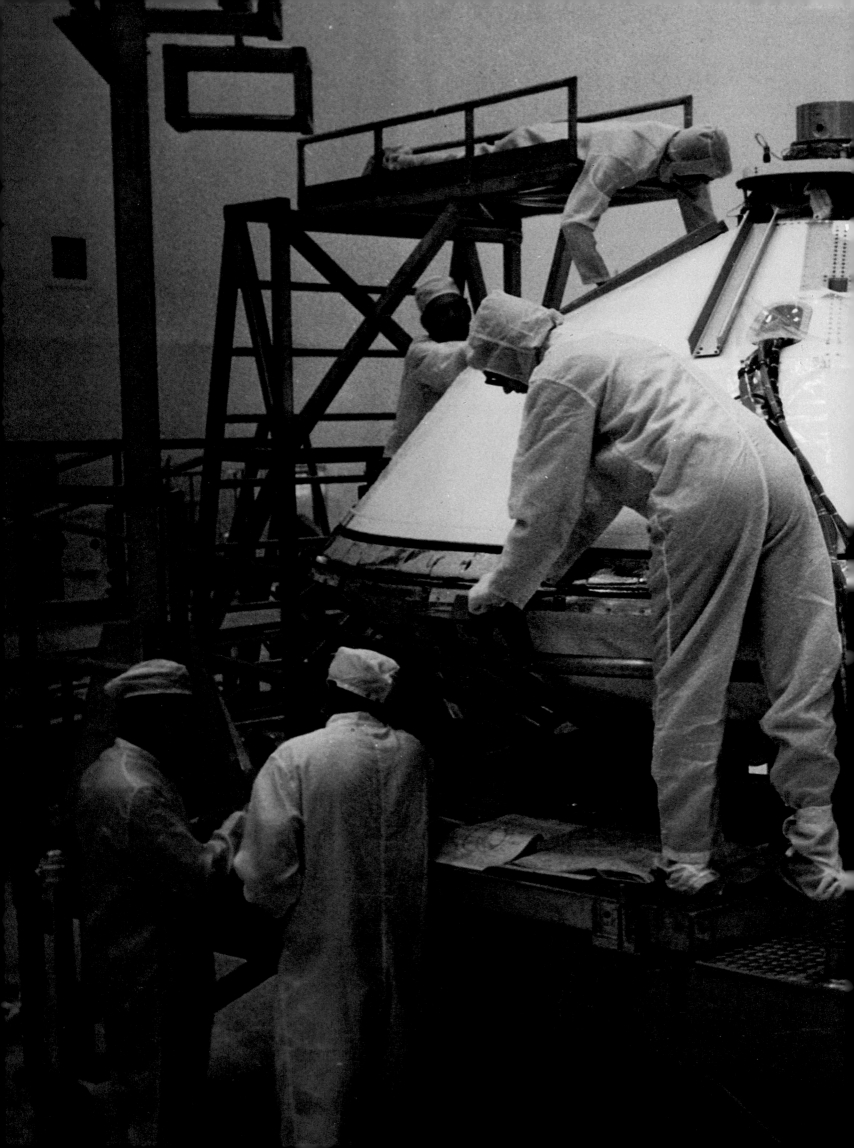

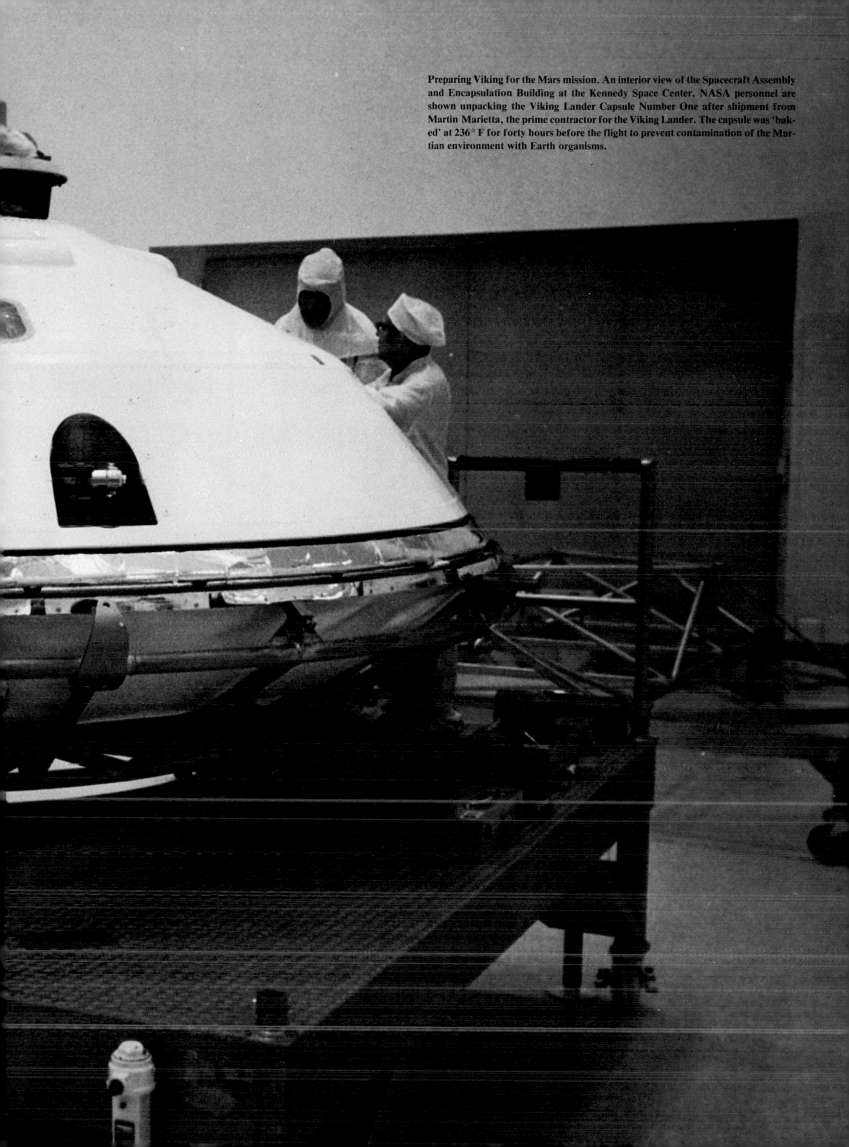

Preparing Viking for the Mars mission. An interior view of the Spacecraft Assembly and Encapsulation Building at the Kennedy Space Center. NASA personnel are shown unpacking the Viking Lander Capsule Number One after shipment from Martin Marietta, the prime contractor for the Viking Lander. The capsule was 'baked' at 236° F for forty hours before the flight to prevent contamination of the Martian environment with Earth organisms.

Viking A

Spacecraft Description—Viking was divided into an orbiter portion and a lander.

The 5125-pound orbiter vaguely resembled a scaled-up Mariner spacecraft, since the arrangement of components was generally similar and Mariner design philosophy was used throughout. The orbiter main structure was a flat, octagonal prism with unequal sides 85 by 99.2 inches along diagonals. Louvers around the periphery of the octagon opened and closed automatically to provide individual thermal control to 16 equipment bays. Four solar panels were hinged at their bases to outrigger structures and hinged again halfway out. Together the panels provided 23,250 square inches of solar cells.

The orbiter's communication system was used as a relay between the lander on the Martian surface and the earth. A parabolic, high-gain antenna, 57.9 inches in diameter, was motor-driven about two axes. A rod-like low-gain antenna on the sunlit side of the spacecraft allowed limited two-way communications with Earth over greater than hemisphere coverage. A third antenna, mounted on the end of one solar panel, was for communication between orbiter and lander.

The orbiter carried three instruments, and performed four experiments. Two narrow-angle television cameras provided high-resolution imaging and was first used for landing-site verification. An atmospheric water detector mapped the super-dry atmosphere of Mars for what water vapor may be there. (Mariner 9 showed signs of water vapor escaping the atmosphere.) An infrared thermal mapper also covered the planet's surface for signs of warmth. And the spacecraft radio provided data for an occultation experiment to provide data on the planet's size, gravity, mass, density and other physical characteristics.

The 2353-pound lander became the first US spacecraft to land on another planet. (US and Soviet spacecraft have landed on the moon, a satellite of the Earth. Soviet spacecraft have landed on Mars and Venus.) The lander was encapsulated in a bioshield before it was launched to keep it from carrying Earth organisms from contaminating Mars—and the biology instrument. Inside the bioshield, which was discarded after reaching the vacuum of space, was an aeroshell that acted as a heat shield during Martian entry.

The lander body was a hollow, six-sided aluminum box 18.2 inches deep enclosed by a top and bottom cover plate. The six sides were 43 inches and 22 inches alternately. The three landing leg units were attached to the shorter sides. Three landing legs were provided instead of four to provide maximum stability.

The lander body provided mounting surfaces on the outside for the two cameras, meteorology boom, surface sampler boom, seismometer, power generators, antennas, descent engines, field tanks, inertial reference unit, various control boxes and the soil inlets for the organic, inorganic and biology instruments. The interior of the body was an environmentally controlled compartment for the biology instrument, gas chromatograph/mass spectrometer, computer, tape recorder, data storage memory, batteries, radios, data acquisition and processing unit and command control and support units.

Landing sequence began when the lander and orbiter were separated and the lander aeroshell made initial contact with

The mating (*left*) of the first Viking Orbiter and the nuclear-powered lander at the NASA launch facility. A mockup (*below*) of a lander showing how it would have explored the surface of Mars. Viking 1 landed at Chryse Planitia in July of 1976 and Viking 2 landed at Utopia in September 1976.

the Martian atmosphere. After atmospheric drag had reduced the spacecraft's speed to 1230 feet per second, an 80-foot parachute was deployed. At 4000 feet altitude the parachute was jettisoned and terminal descent engines reduced velocity to only a few feet per second. These engines shut down just above the Martian surface. There were three main terminal descent engine clusters, each containing 18 tiny nozzles. Engineers adopted this design to spread exhaust gases from ultrapure hydrazine to minimize the possibility of surface contamination.

Project Objectives—To conduct a systematic investigation of Mars from orbit and from the surface, with primary emphasis on the search for life.

Spacecraft Payload (Viking 1)—Three instruments and the spacecraft radio on the orbiter and nine instruments on the lander; Orbiter experiments—high-resolution camera, atmospheric water-vapor mapper, surface heat mapper, occultation experiment. Lander experiments—biology instrument, gas chromatograph/mass spectrometer, X-ray fluorescence spectrometer, seismometer, meteorology instrument, stereo color cameras, physical and magnetic properties of soil, aerodynamic properties and composition of Martian atmosphere with changes in altitude.

Spacecraft Payload (Viking 2)—Three instruments and the spacecraft radio on the orbiter and nine instruments on the lander. Orbiter experiments—high-resolution camera, atmospheric water-vapor mapper, surface heat mapper, occultation experiment. Lander experiments—biology instrument, gas chromatograph/mass spectrometer, X-ray fluorescence spectrometer, seismometer, meteorology instrument, stereo color cameras, physical and magnetic properties of soil, aerodynamic properties and composition of Martian atmosphere with changes in altitude.

Project Results—Viking 1 was launched from Kennedy Space Center on 20 August 1975; it successfully soft-landed on 20 July 1976.

Project Results—Viking 2 was launched 9 September 1975, successfully soft-landed on Mars 3 September 1976.

Preparing Viking for the epic voyage to Mars. Boeing engineers monitoring electromagnetic interference (EMI) testing of the Viking satellite platform. Inside an anechoic chamber at the Boeing Space Center at Kent, Washington, the satellite platform's electrical systems were tested against each other to ensure there was no interference during normal spacecraft operations. The EMI testing of such items as the power and telemetry ground data systems verified that there were no false signals during electronic sensoring of Earth and Sun. A technician (*above, top*) installs propellant tanks during buildup of the Viking Lander Capsule No 1 in the Spacecraft Assembly and Encapsulation Building at the Kennedy Space Center. The tanks are part of the terminal descent propulsion system. An interior view (*above*) of the Spacecraft Assembly and Encapsulation Building. The first object of all this preparation landed Viking on the Martian desert Chryse Planitia on 20 July 1976.

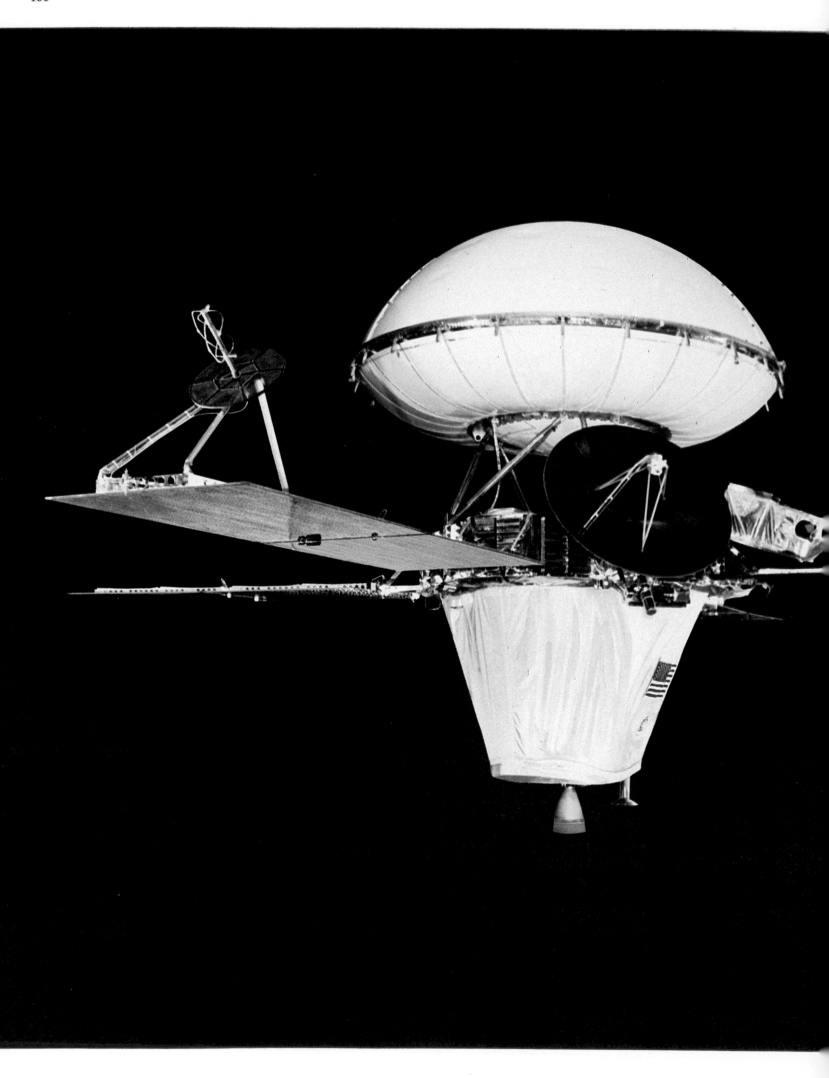

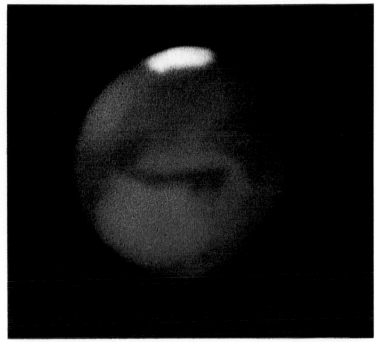

The Viking mission to Mars. Two Viking spacecraft (*facing page*), each consisting of an orbiter and a lander, were launched from the Kennedy Space Center in Florida (*above*). Viking 1 was launched in August of 1975 and Viking 2 lifted off the pad one month later. The objective of the Viking mission was to study the planet Mars. This effort included evaluation of the Martian environment from orbit, the determination of the composition of the planet through soil analysis, and the search for life on Mars. The fanciful observations of Martian canals by Percival Lowell would soon be forever put to rest. The complexion of Mars as viewed through ground based telescopes (*above, top*) would soon be understood. The Viking mission was the first attempt by the United States to land a spacecraft on another planet. Viking 1 touched down in July of 1976 and Viking 2 landed in September of the same year.

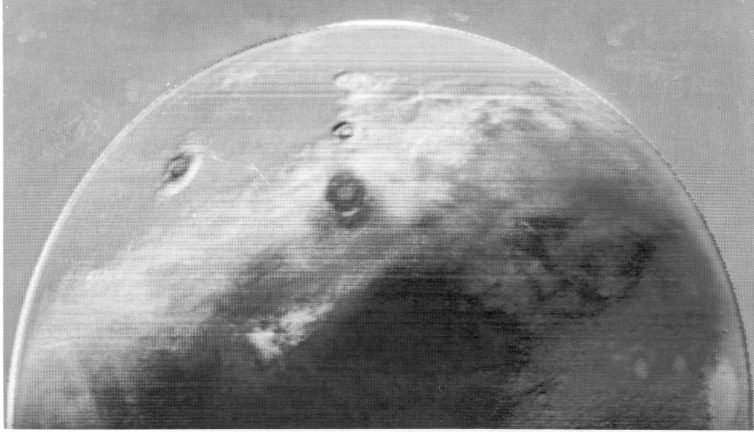

Enroute to Mars! The Viking spacecraft returned some spectacular images of Mars while traveling towards the planet. Viking 2 returned this dramatic color photograph (*above, top*) while still more than 260,000 miles away. Viking 2 approached Mars more from the dark side than had Viking 1. This approach allowed the stunning crescent view of the planet. Computer enchanced images (*above, bottom*) of the planet helped planetary scientists to learn about Mars. This image was used to separate and amplify extremely subtle color differences among various clouds, atmospheric haze, surface frosts and rock materials. The violet background is an artifact representing the black of space and the Martian night. Relatively bright materials, atmospheric hazes, surface frosts and bright deserts are represented by turquoise, whites and yellows. Darker materials on the surface are represented by deep reds and blues. The green area is an overexposure due to the tremendous range of brightness across the planet. The giant Martian volcanoes, which are dark red, are both redder and darker than the plains materials, shown by the yellow-orange colors, which fill the areas around them. A broad band of atmospheric haze, colored bluish-white extends across the volcanoes toward the northern edge of the planet. As the orbits moved closer to the planet, stark views of the planet emerged (*facing page*). This orbiter photograph reveals the thin carbon dioxide atmosphere that envelopes Mars. Several high altitude cloud layers are visible on the horizon. The large basin at left center, named Argyre Planitia, is the scar of an ancient asteroid.

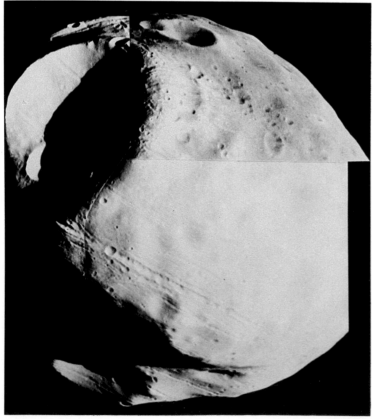

In all, the Viking cameras transmitted over 52,000 photographs of Mars and its surrounding environment. This pictorial history has given planetary scientists a wealth of data from which to learn about Mars. Some 4500 of these photographs from the surface depict the Martian landscape through the changing seasons. It has been estimated that these photographs have expanded our knowledge of the geology of Mars by at least 20 times over what was previously known. As a result, the spectacular volcanoes and deeply carved canyons that characterize the planet are now familiar features to scientists and non-scientists alike.

A most interesting Martian volcano (*above and facing page*) towers over the surrounding plains. At its highest, Olympus Mons is over sixteen miles high (about 84,500 feet). This compares to Hawaii's Mauna Loa, the largest Earth volcano, which rises a mere five miles above the floor of the ocean. The summit caldera of Olympus Mons comprises a series of craters formed over the centuries by innumerable eruptions, crater collapses and more eruptions. Great lava flows extend from the volcano for a thousand miles or more. Although not clearly seen in these images, enormous canyons and cliffs are evidence of Marsquakes in the past. Based on data transmitted from Viking it appears that the planet's volcano and quake activities are nonexistent now (or at least temporarily inactive). As fascinating as the surface of the planet were the satellites of Mars.

Phobos (*left*) was once the subject of heated debate among nineteenth century astronomers. This fearsome moon was thought to be an artificial satellite launched by the Martians. This photomosaic of Phobos shows the front side of the moon, which always faces Mars. Stickney, the largest crater on Phobos, is six miles across. The linear grooves in the center passing through Stickney appear to be fractures in the surface caused by the impact which formed the crater. Kepler Ridge in the southern hemisphere is seen casting a shadow (over the large crater, Hall, at the bottom). It is the peculiar shape and textural surface that have prompted some science writers to describe Phobos as looking 'somewhat like a potato.' Deimos, Mars' other moon, appears smoother and more spherical by comparison.

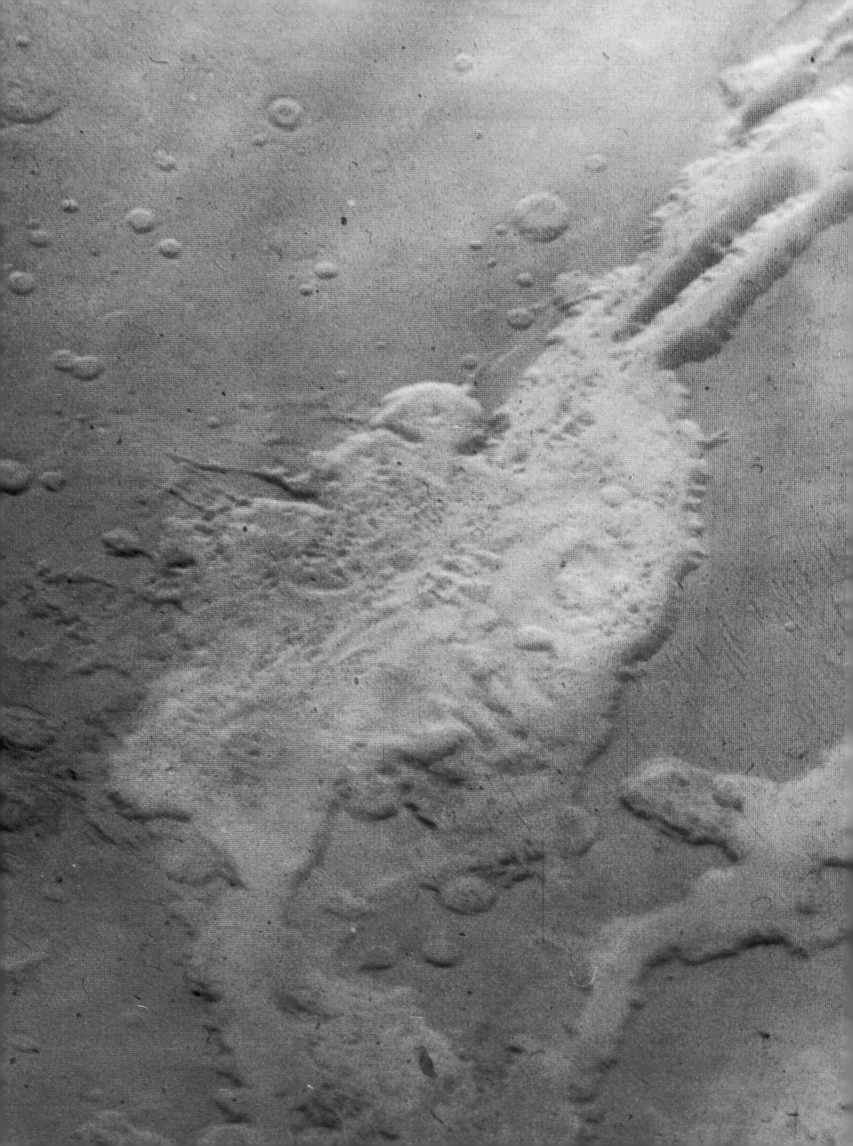

This photograph of Mars was taken by Viking Orbiter 1. Valles Marineris, as long as the North American continent from coast to coast, stretches across the surface. The surface is marred by the scars of meteor strikes.

One of the first pictures taken on the surface of Mars by the Viking 1 Lander shows that the soil consists mainly of reddish, fine-grained material. Most of the rocks are coated with a reddish stain except where the rock surface has been freshly fractured or abraded. There is a group of rocks near the horizon which appear to be free of the reddish stain. These may be relatively young volcanic rocks or older rocks recently excavated from the subsurface.

Martian vistas as seen from the Viking landers at work on the surface of the planet. Most of the rocks (*facing page, top*) measure about 20 inches across except for 'Big Joe' at left which is about 8 feet wide. The Lander scoop for collecting soil samples (*facing page, bottom*) is shown about to gather a sample. Also shown is the brush used to clean off the scoop. Noon time on Mars (*above, top*) as seen from the two cameras onboard Viking 1. This view was taken so that scientists might use a stereo viewer to study the planet. The historic first photograph from the surface of Mars (*above*) showing the foot of the Lander. The flag of the United States (*right*) with the rocky Martian surface in the background. Also shown is the logo of the American bicentennial that Viking 1 carried to Mars. This picture was taken at local Mars time of 7:18 am, hence the relatively dark sky.

Summer and winter seasons on Mars can be clearly seen from the images of arid rock-strewn plains (*facing page, top*) looking south from the lander. Many of the rocks shown, some as large as three feet across, are porous and spongelike, similar to some of Earth's volcanic rocks. Other rocks are coarse-grained such as the large rock at lower left. Between the rocks the surface is blanketed with fine-grained material that tends to pile into drifts. A similar Martian view (*facing page, bottom*) shows the surface covered by an extremely thin blanket of ice. Scientists believe that dust particles in the atmosphere pick up bits of water and combine with carbon dioxide (which makes up about 95 percent of the atmosphere), freezes and falls to the surface like snow or frost. Once on the surface, the Sun evaporates the carbon dioxide from the frost, leaving behind the water and dust which remain on the surface during the Martian winter, about 100 days. The winter period occurs every Martian year which corresponds to about twenty-three Earth months. The ice seen in this picture is extremely thin, perhaps no more than one-thousandth of an inch thick. The surface of the planet is mostly covered by a layer of reddish dust. Subsurface soil is a darker material (*above*) as can be seen in the dark debris, identified by the arrow, kicked up by the Viking 1 Lander. The dust visible atop the Lander footpad was stirred up during the landing of the craft. This sort of dust provided scientists with a means of viewing apparent wind on Mars without the use of onboard instrumentaion. Computers aboard the Viking Lander had a computer imaging capability (*overleaf*) which afforded planetary scientists wide angle vistas of the Martian landscape. Pictures of the landing site showed no evidence of plant life, and a gas-chromatographer detected no complex organic molecules on Mars. This, of course, doesn't rule out the possibility that some form of life might one day be found.

VIKING LANDER 2   CAMERA 2
DIODE BELYT   STEP SIZE 0.1
VIKING LANDER 2   CAMERA 2
DIODE BELYT   STEP SIZE 0.1
VIKING LANDER 2   CAMERA 2
DIODE BELYT   STEP SIZE 0.1
COLOR MOSAIC OF RADCAM OUTPUT   SPE
LABCAT
SAR - LGEDM
MASKVL

                              SEGMENT 1 OF

CE LABEL 22A003/000
CHANNEL MODE 2/1
CE LABEL 22A016/002
CHANNEL MODE 2/1
CE LABEL 22A018/002
CHANNEL MODE 2/1
IN 0. MAX 4.5

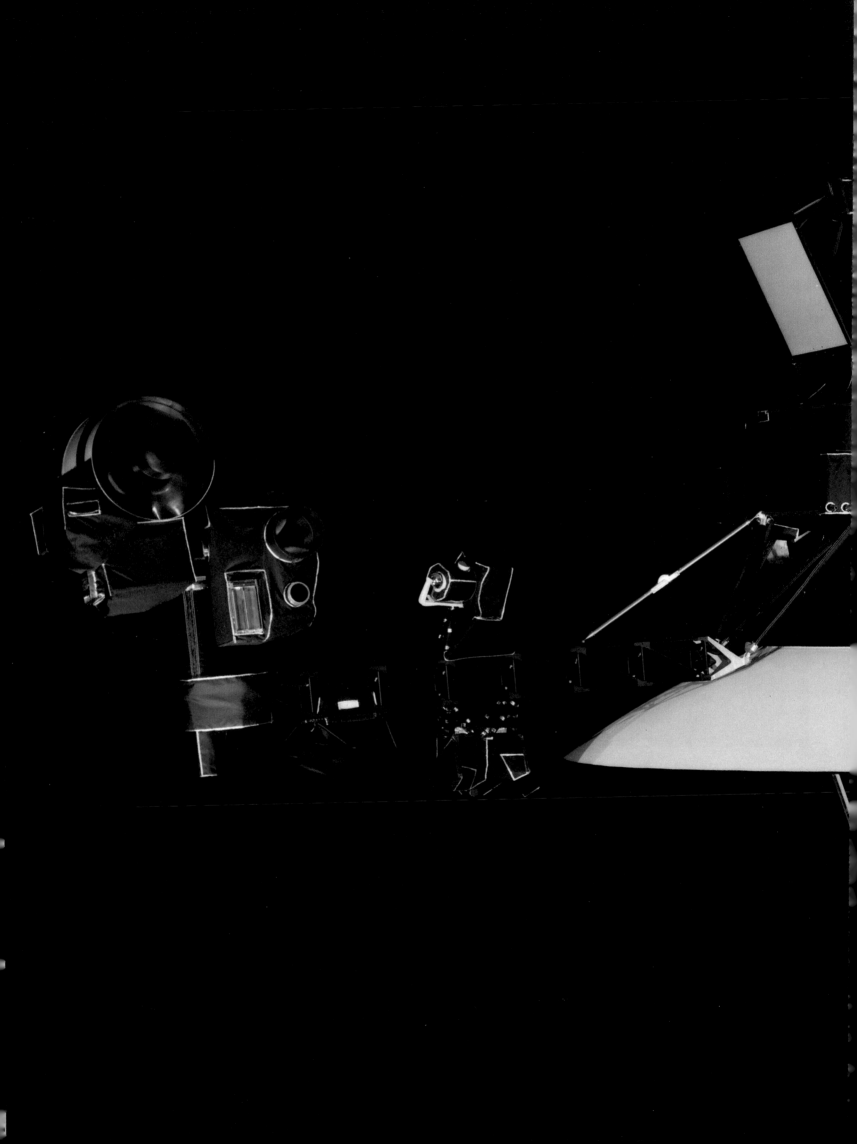

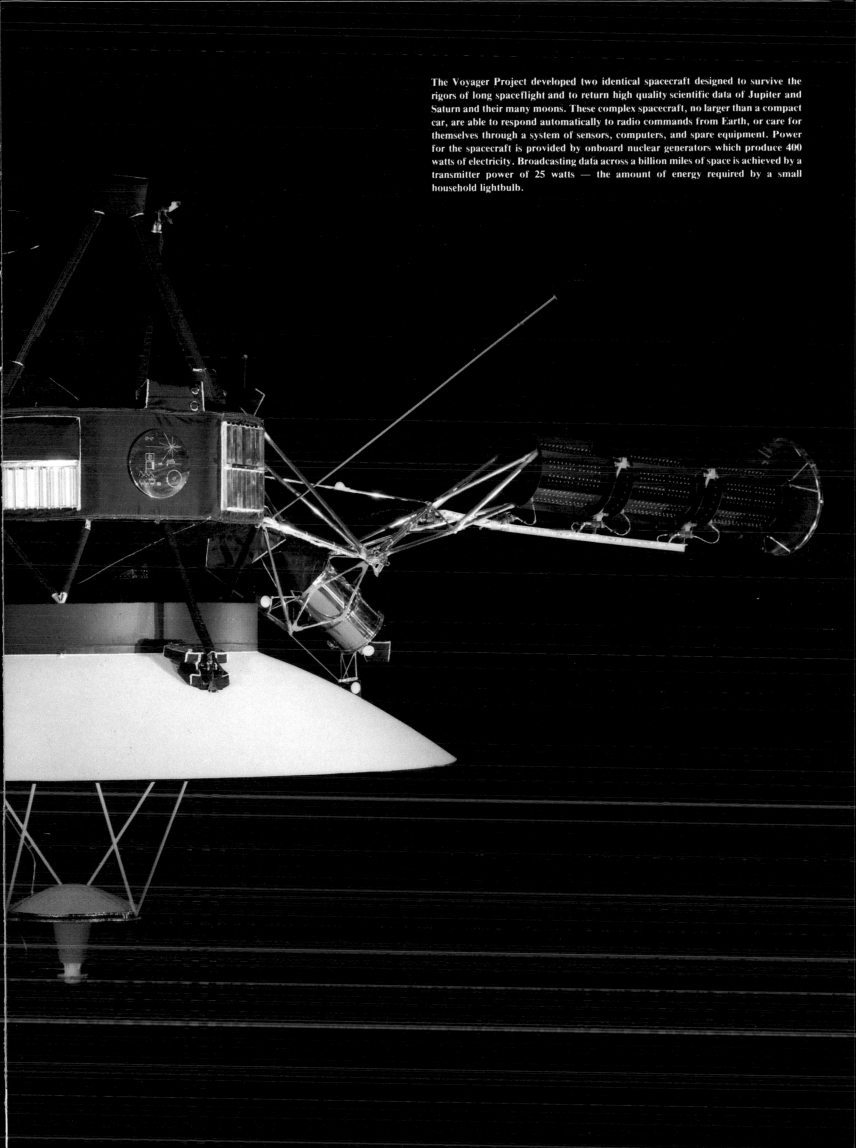

The Voyager Project developed two identical spacecraft designed to survive the rigors of long spaceflight and to return high quality scientific data of Jupiter and Saturn and their many moons. These complex spacecraft, no larger than a compact car, are able to respond automatically to radio commands from Earth, or care for themselves through a system of sensors, computers, and spare equipment. Power for the spacecraft is provided by onboard nuclear generators which produce 400 watts of electricity. Broadcasting data across a billion miles of space is achieved by a transmitter power of 25 watts — the amount of energy required by a small household lightbulb.

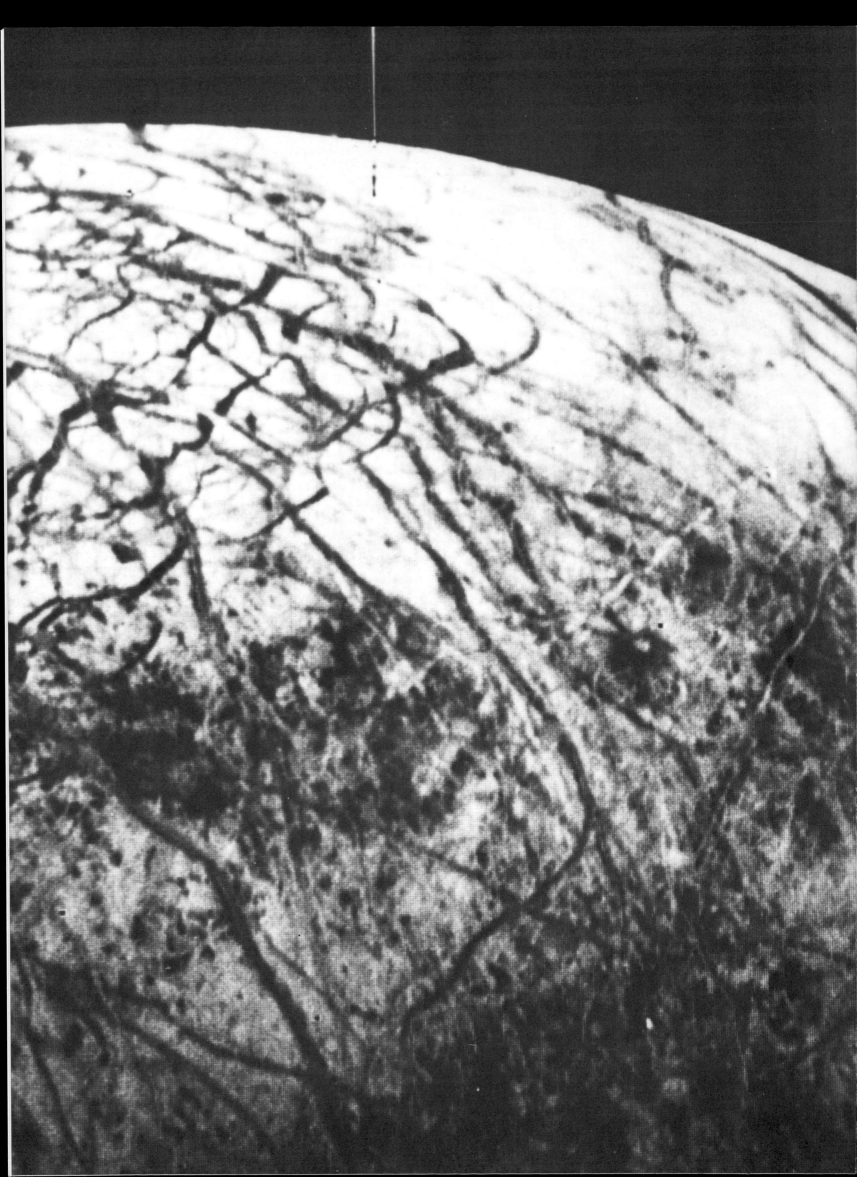

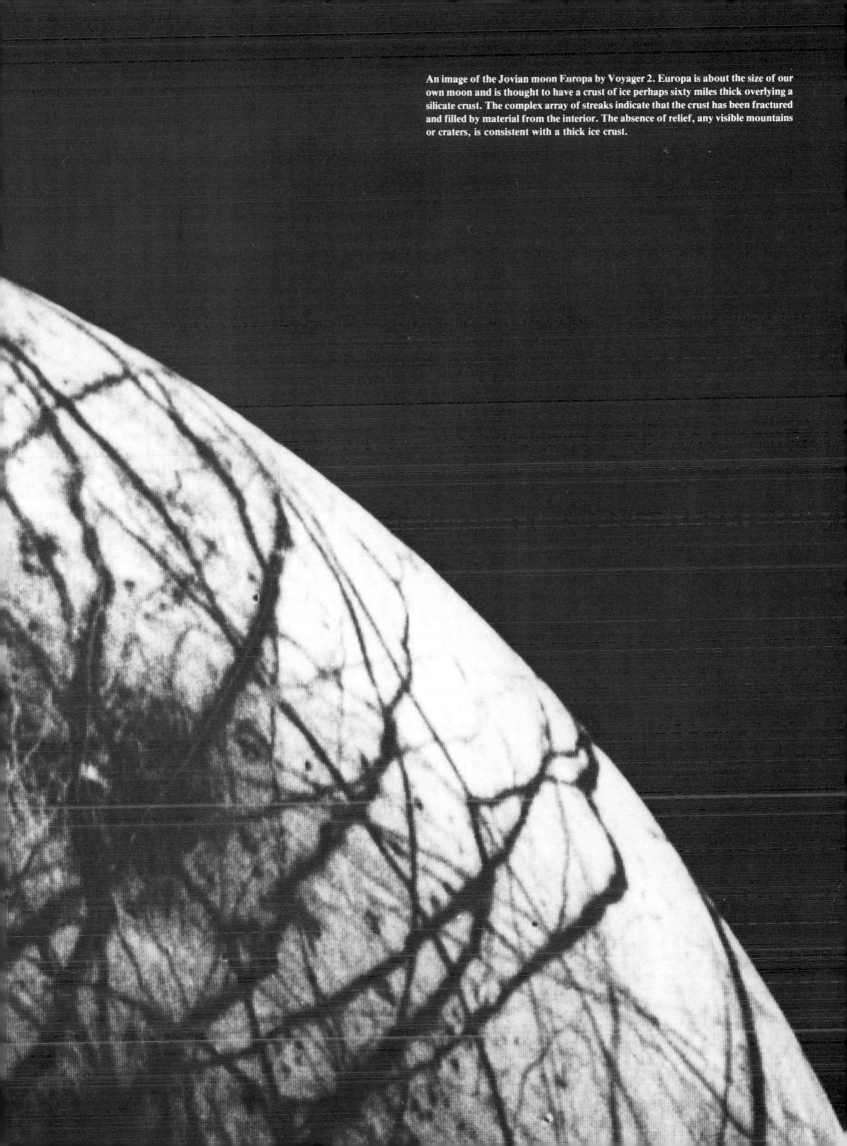

An image of the Jovian moon Europa by Voyager 2. Europa is about the size of our own moon and is thought to have a crust of ice perhaps sixty miles thick overlying a silicate crust. The complex array of streaks indicate that the crust has been fractured and filled by material from the interior. The absence of relief, any visible mountains or craters, is consistent with a thick ice crust.

130

A composite of Jupiter and its four planet-sized moons called the Galilean satellites: Io (*upper left*); Europa (*center*), Ganymede, and Callisto (*right*). Io's surface is composed of sodium salts and sulfur, Europa's surface is mostly water ice, Ganymede's surface is a mixture of rock and water ice, and Callisto's surface resembles an enormous ball of frozen slush.

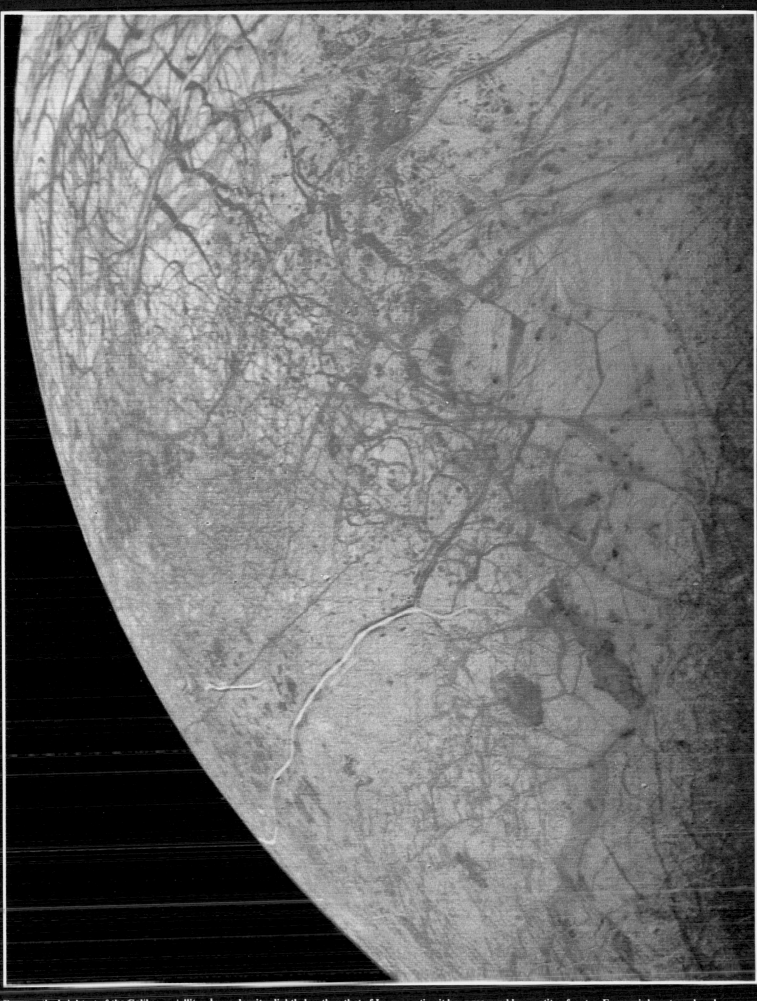

Europa, the brightest of the Galilean satellites, has a density slightly less than that of Io, suggesting it has a reasonable quantity of water. Europa is less strongly colored than Io and is probably the smoothest object in the solar system — it may be covered with a planet-wide glacier of flowing water ice that tends to erase the evidence of meteoritic impacts.

A Voyager 2 image of Ganymede from a distance of 195,000 miles showing features down to about four miles across. The broad light regions running through the image are the typical grooved structures seen within most of the light regions of the satellite. To the lower left is an example of what might be evidence of large scale lateral motion of Ganymede's crust. The band of grooved terrain in this region appears to be offset by more than 30 miles. These are the first clear examples of strike-slip style faulting on any planet other than Earth.

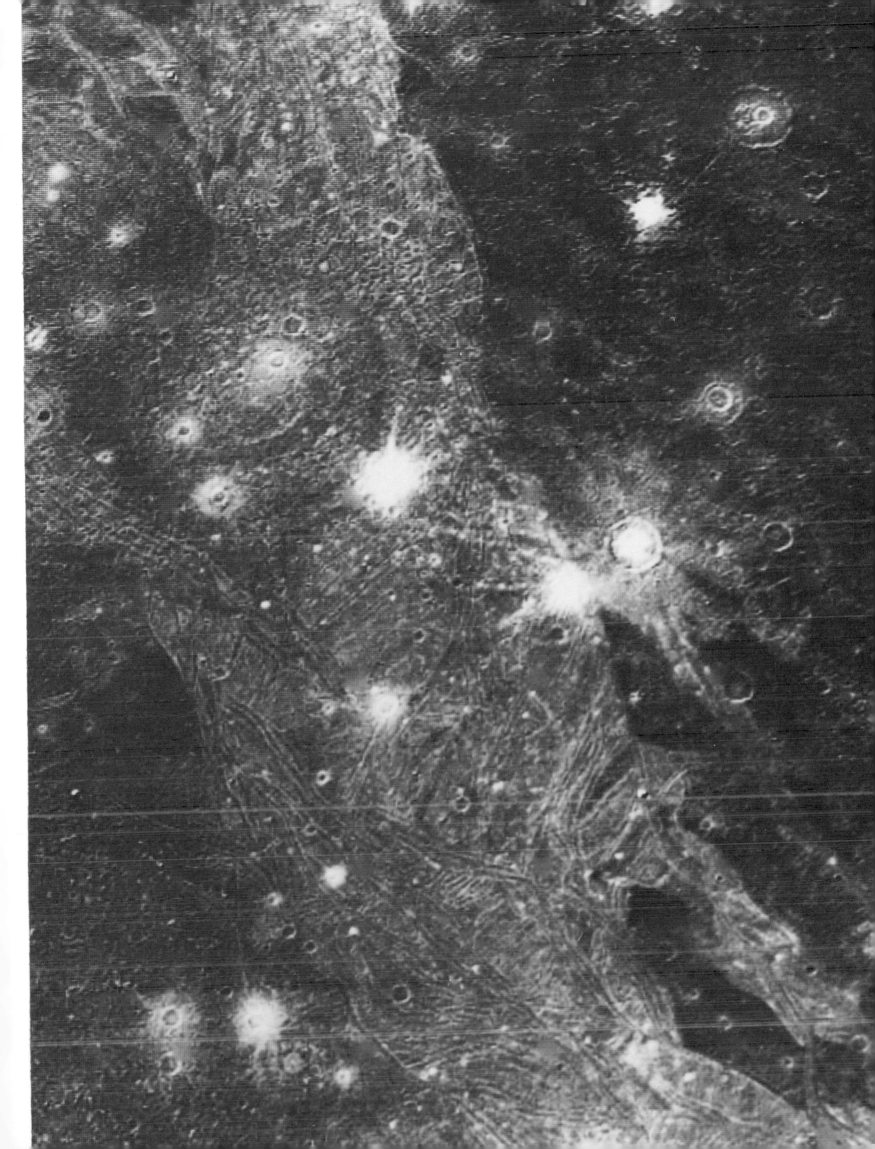

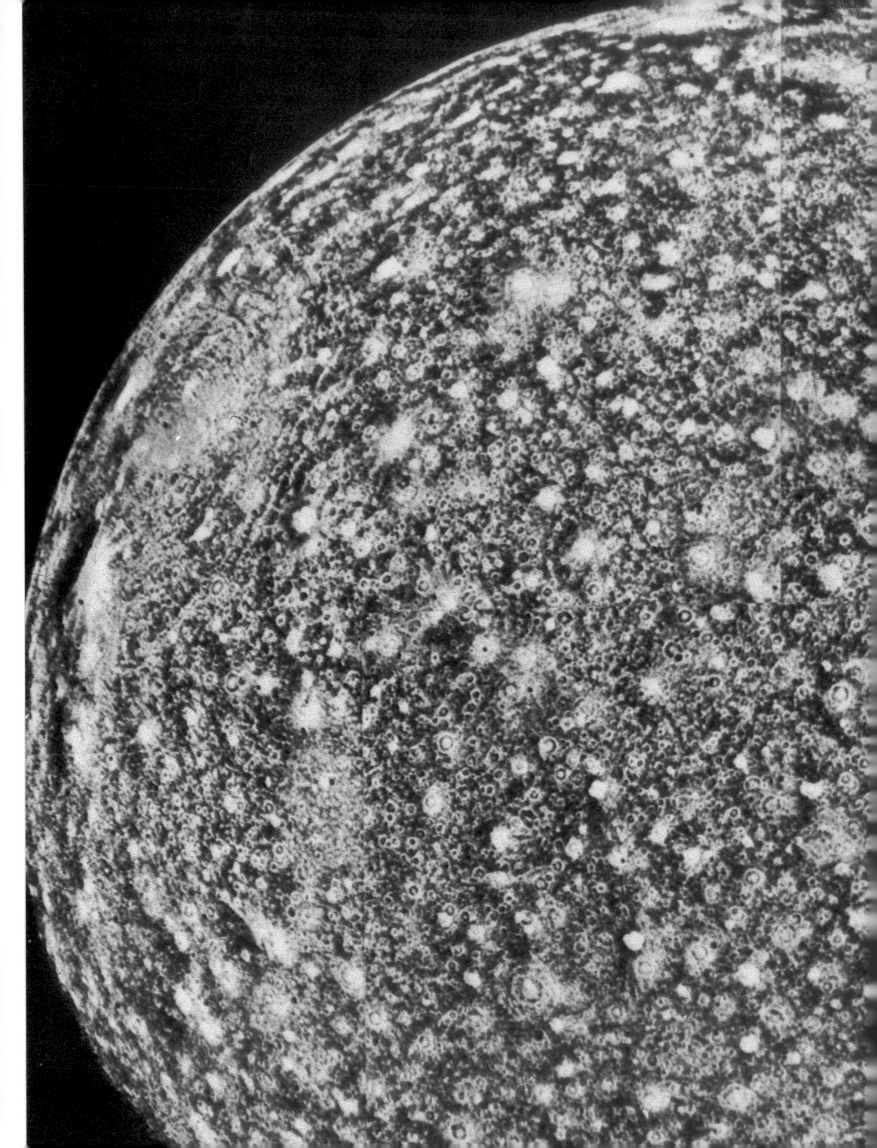

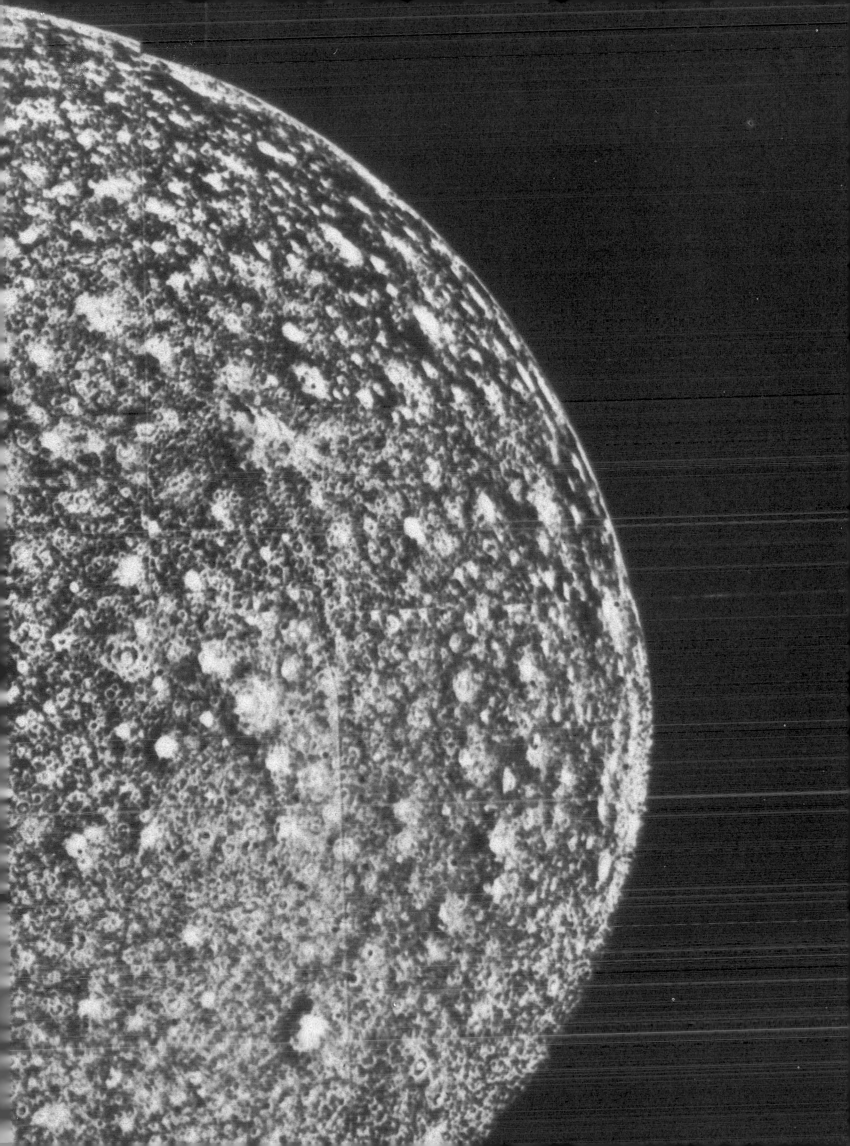

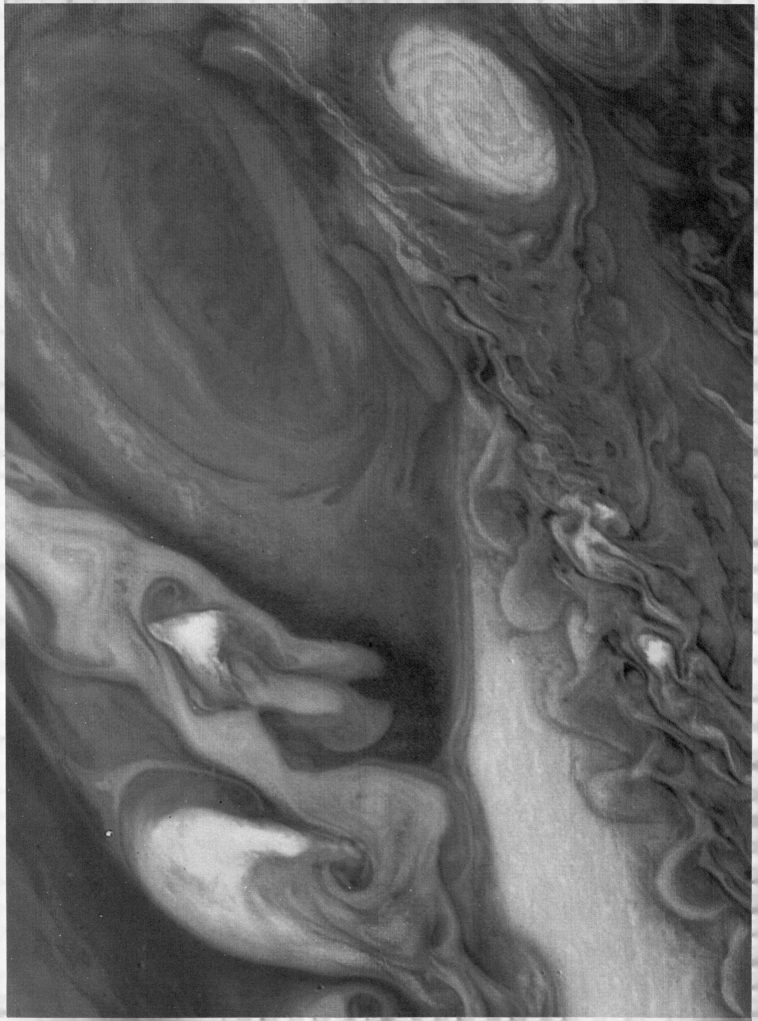

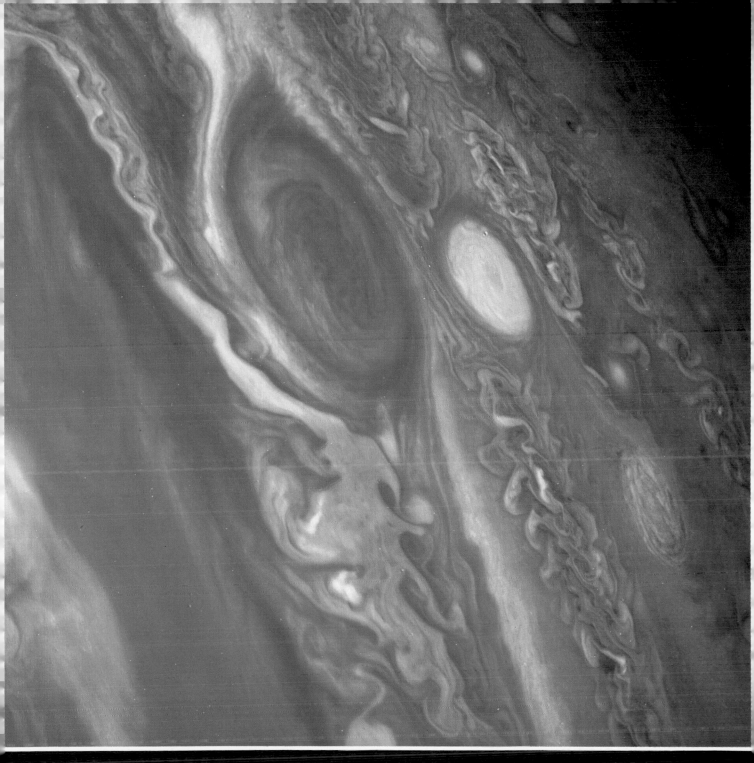

The atmosphere of Jupiter is complex, with bands of alternating colors. The turbulent atmosphere appears to be controlled by forces far below the visible cloud tops. The Great Red Spot (*left and above*) is a tremendous atmospheric storm that rotates counterclockwise with one revolution every six days. Generally dark features on the planet (*right and far right*) are warmer than the lighter areas. The exception is the Great Red Spot, which is the coldest place on the planet and covers an area of about 30,000 miles long and 7,000 miles wide, about three times the surface area of the Earth. It is believed to soar about 15 miles above the surrounding clouds. The dominant large scale motions in the Jovian atmosphere are west to east with small scale eddy-like circulation between the bands. A very thin ring of material circles the planet above the cloud tops. Auroras and cloud-top lighting bolts, similar to superbolts on Earth, were observed and photographed by Voyage.

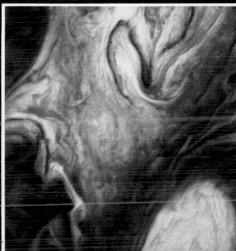

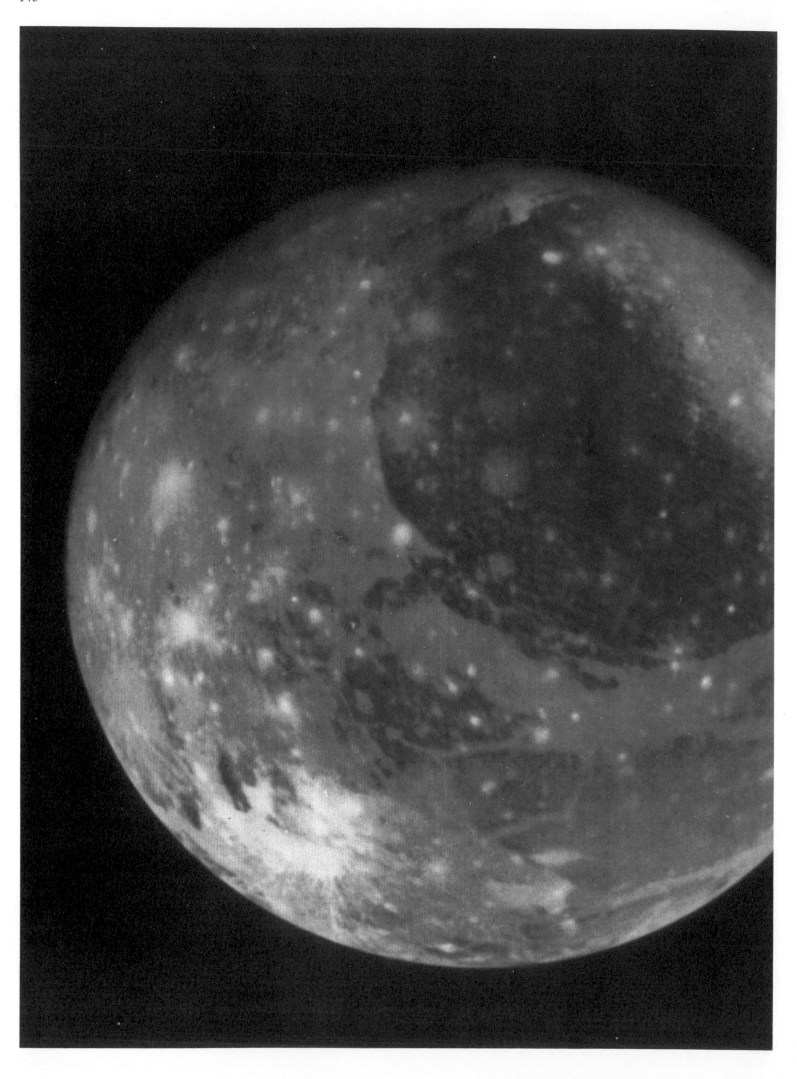

Ganymede (*facing page*) is Jupiter's largest satellite, has diameter of about 1.5 times that of our Moon and is larger than the planet Mercury. Ganymede has a density only twice that of water and is composed of a mixture of rock and water ice. The grooves may indicate recent activity resembling continental drift on Earth.

The bright spots dotting the surface of Ganymede (*above*) are relatively recent impact craters. The light branching bands are ridged and grooved terrain first seen by Voyager 1 and are younger than the more heavily cratered dark regions. These are probably the internal and external responses of the icy crust of the satellite.

Voyager was able to image a small ring system around Jupiter as shown in this photograph of the ring illuminated by sunlight coming from behind the planet.

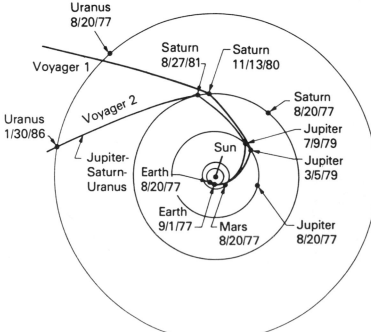

Uranus
8/20/77

Voyager 1

Saturn
8/27/81

Saturn
11/13/80

Voyager 2

Saturn
8/20/77

Uranus
1/30/86

Jupiter
7/9/79

Jupiter-
Saturn-
Uranus

Sun

Jupiter
3/5/79

Earth
8/20/77

Earth
9/1/77

Mars
8/20/77

Jupiter
8/20/77

The path (*left*) of the two Voyager spacecraft took them near Mars, Jupiter, Saturn and Uranus. The slightly different encounter dates of the two craft with solar system objects gave planetary scientists the opportunity to compare differences in objects and proved invaluable in learning about the environments of our sister planets. Voyager 2's mission to Uranus returned thousands of images and voluminous amounts of other data on the planet, its system of rings and its ten moons. Voyager's next encounter, scheduled for August 1989, will be with Neptune.

Io, Jupiter's largest innermost moon (*above and right*), has a surface composed of sodium salts and sulfur and is distinguished by its bright orange surface. Voyager has shown this moon to have active volcanic activity. These two dramatic photographs clearly show plumes of volcanic material spewing 60 to 100 miles into the atmosphere from enormous volcanic explosions. An eruption of this magnitude would require an ejection velocity of about 1200 miles per hour. As a result, material would reach the crest of the fountain in several minutes. Volcanic explosions similar to this occur on Earth when magma gasses expand explosively as material is vented.

On Earth, water is the major gas driving a volcanic explosion. Because Io is thought to be extremely dry, scientists are searching for other gases to explain the explosion. In addition to the plumes, a variety of features can be seen which appear linked to the intense volcanic center. At least two general types of volcanic activity appear on Io: Explosive eruptions which spew material into the sky as seen here and lava flows that vent from the surface. These lava flows may be emanating from the dark spot (*overleaf*) which may be a volcanic crater. Io is the first body in the solar system (other than Earth) where active volcanism has been observed.

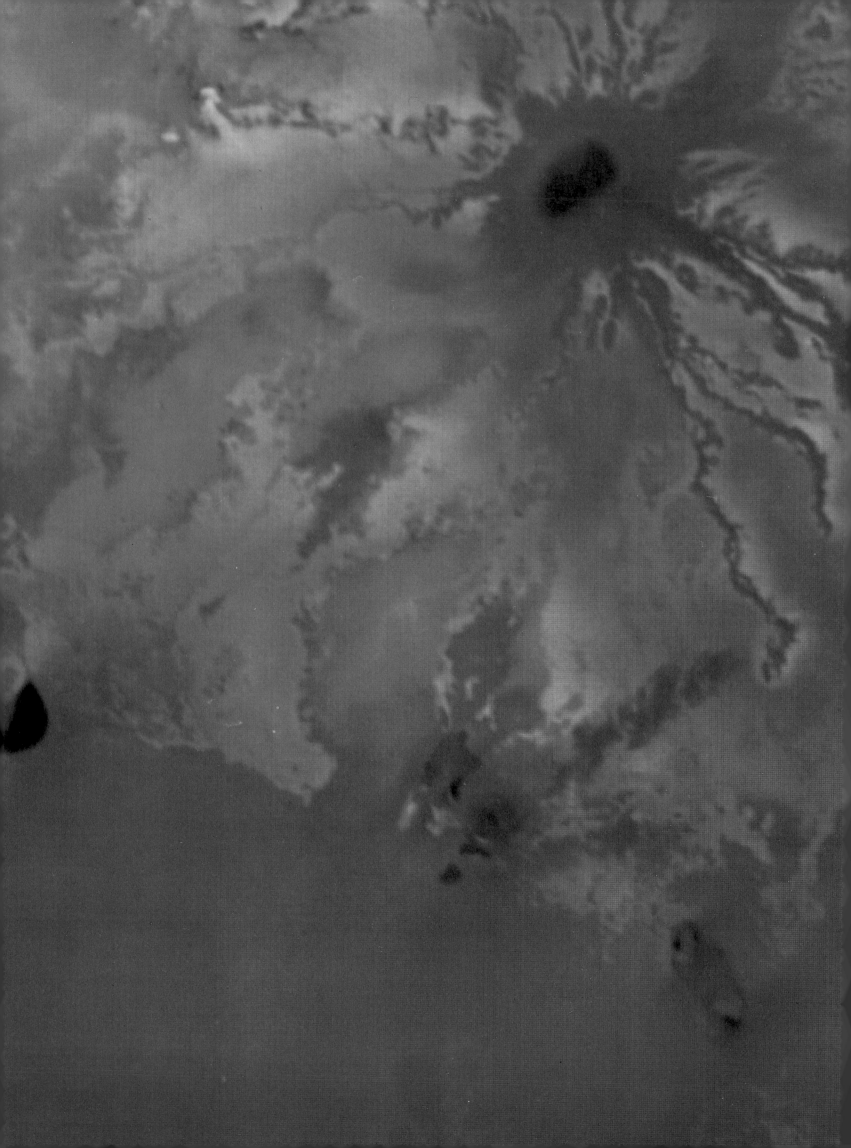

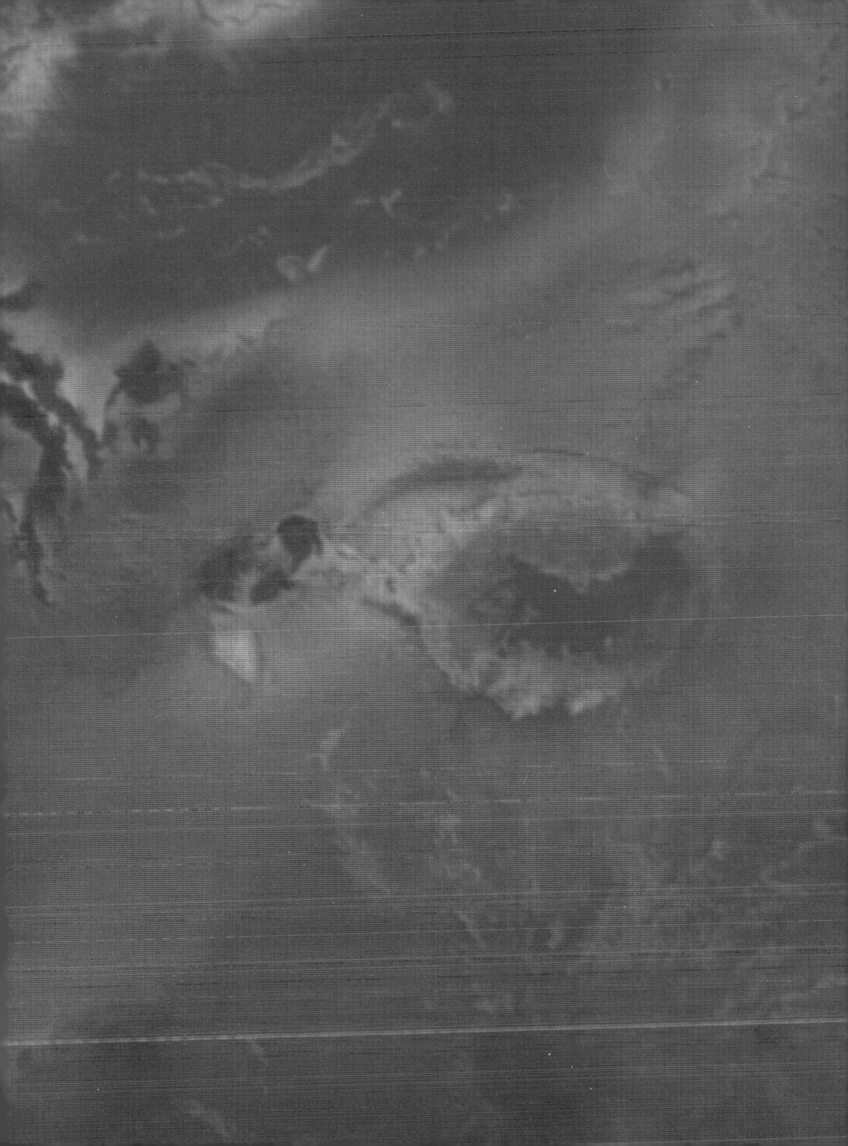

This true-color image of Saturn is a composite of Voyager images from a distance of 13 million miles. Three of Saturn's icy moons are visible at top left. They are, in order of distance from the planet: Tethys, 652 miles in diameter; Dione, 696 miles in diameter; and Rhea, 951 miles in diameter. The pastel and yellow hues on the planet reveal many contrasting bright and darker bands in both hemispheres of Saturn's weather system.

Further study with computer-enhanced coloration increase the visibility of large, bright features in the North Temperate Belt. It is believed that these spots might resemble gigantic convective storms (similar to but much larger than thunderstorms in the Earth's atmosphere) with upwelling from deep within Saturn's atmosphere.

The trajectories of the two Voyager spacecraft differed in many ways even though both were launched in the same year from Kennedy Space Center. Part of the difference was the requirement that Voyager 2 continue onto Uranus and Neptune. In addition, efforts were directed toward making the second encounter as complementary as possible to the first.

Voyager 2 approached Saturn (*below*) looking down on the northern lighted side of the rings and crossed the ring plane only once, very close to the time of its closest passage by Saturn. The spacecraft exited from the Saturn system looking back at the south or dark side of the rings. This trajectory provided for the close flybys of the satellites Iapetus, Hyperion, Enceladus, and Tethys. Both Voyagers are expected to exit the heliosphere (the outer edge of the solar wind) in the 1990s.

The Saturn satellite Tethys (*left and facing page* with the satellite Dione) is one of a class of intermediate-sized bodies of the Saturn system which have never before been studied by spacecraft. These intermediate-sized satellites include Mimas, Enceladus, Dione, and Rhea, in addition to Tethys. These moons are all icy bodies with surfaces which are scarred, excepting Enceladus which is smooth despite eons of meteoric bombardment. Internal processes may have erased the evidence of early bombardment from the surface of Enceladus.

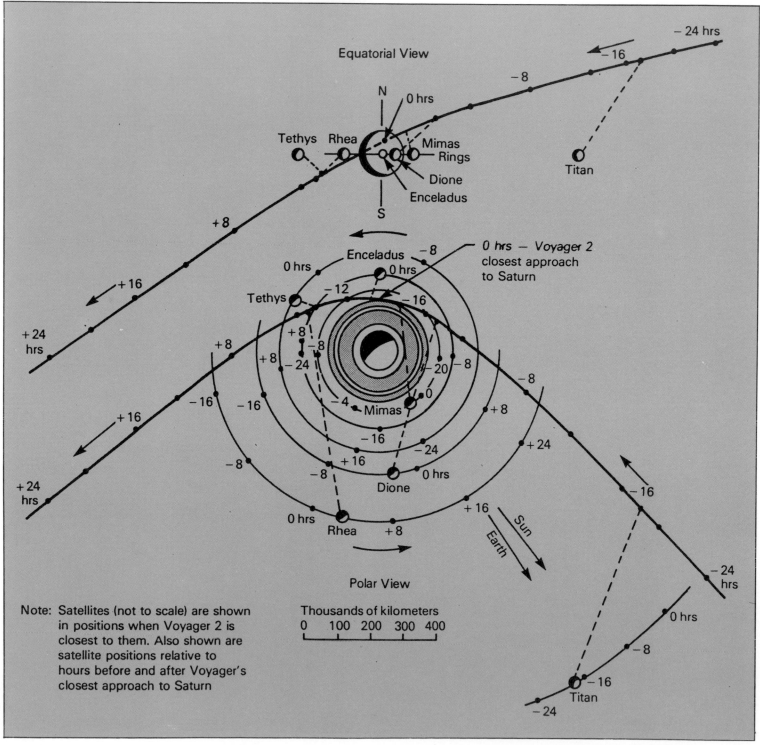

Note: Satellites (not to scale) are shown in positions when Voyager 2 is closest to them. Also shown are satellite positions relative to hours before and after Voyager's closest approach to Saturn

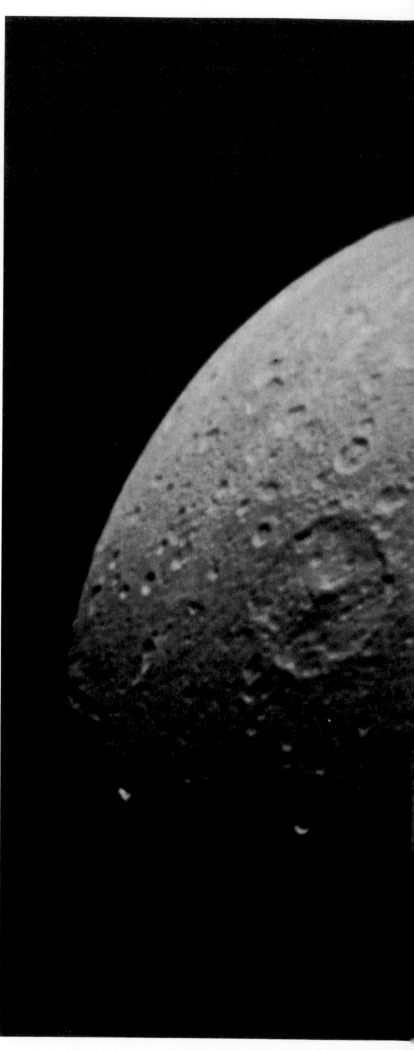

Titan (*above, top*) is remarkable for many reasons. It is the second largest moon in the solar system (Ganymede of Jupiter system is slightly larger). It is the only moon known to have a thick atmosphere, perhaps 10 times as thick as that of Earth. Note the brighter clouds in the southern hemisphere and the dark hood near the north pole. These differences in shading may be due to seasonal variations and the influence of Saturn's magnetic field. In many ways the surface of the moon Enceladus (*above, bottom*) resembles that of Jupiter's Ganymede. Linear sets of grooves miles long traverse the surface and are probably faults resulting from deformation of the crust. The cratered regions are geologically young and suggest that the moon has experienced recent internal melting. The icy surface of Dione (*right*) is scarred by the infall of cosmic debris. Sinuous valleys are probably the result of crustal fracturing. Dione is much smaller than any of Jupiter's moons.

Titan is the only satellite known to have a thick atmosphere. Titan's atmosphere is composed mainly of nitrogen, with smaller amounts of methane and other hydrocarbons, and may be ten times as dense as the atmosphere of Earth. The entire moon is surrounded by a haze which obscures Titan's solid surface. The temperature at the surface of Titan, which is probably near -280° F, may allow for the formation of lakes and rivers of liquid methane.

The complex ring system of Saturn from a range of 445,000 miles. Voyager discoveries included the braided F ring, dark spokelike features in the B ring, and a complex structure of ringlets in the Cassini division. Rings are visible because they reflect the light of the sun.

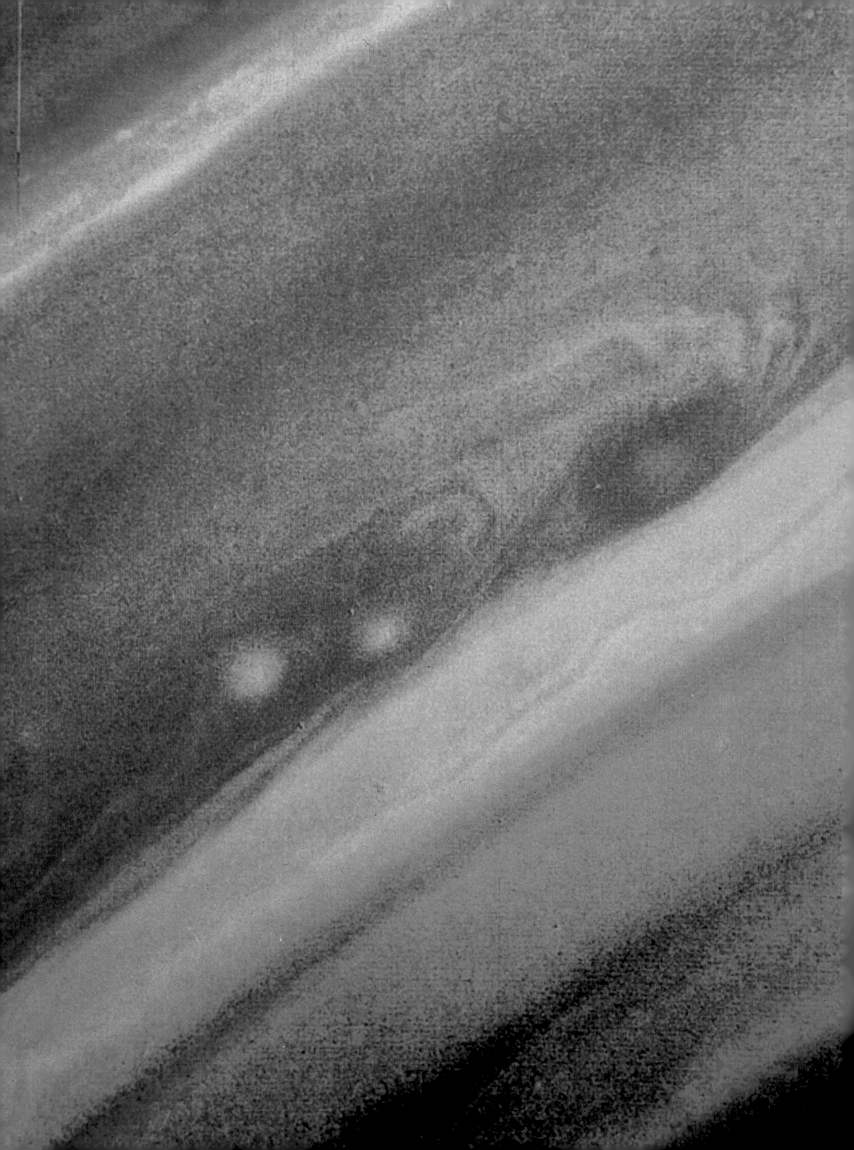

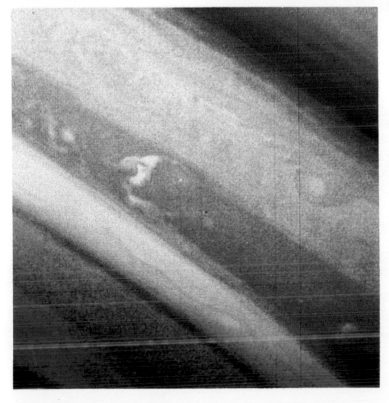

The complexity of the Saturn system is seen in this series of images. The powerful weather systems are seen from a perspective of the northern hemisphere (*facing page*) and a color enhanced image (*left*) of the clouds of Saturn. Saturn also has a ring system (*above*), which has intrigued astronomers since the days of Galileo. Saturn and its six largest moons are grouped in an artist's montage (*above, top*) assembled from images taken by Voyager 1 in November 1980. Clockwise, starting from the far right, the satellites are Tethys and Mimas (in front of the planet), Enceladus (in front of the rings), Dione (in the left forefront), Rhea (off the left edge of the rings), and Titan (at top). Studies show that Titan's atmosphere is abundant in nitrogen, not methane as was earlier predicted.

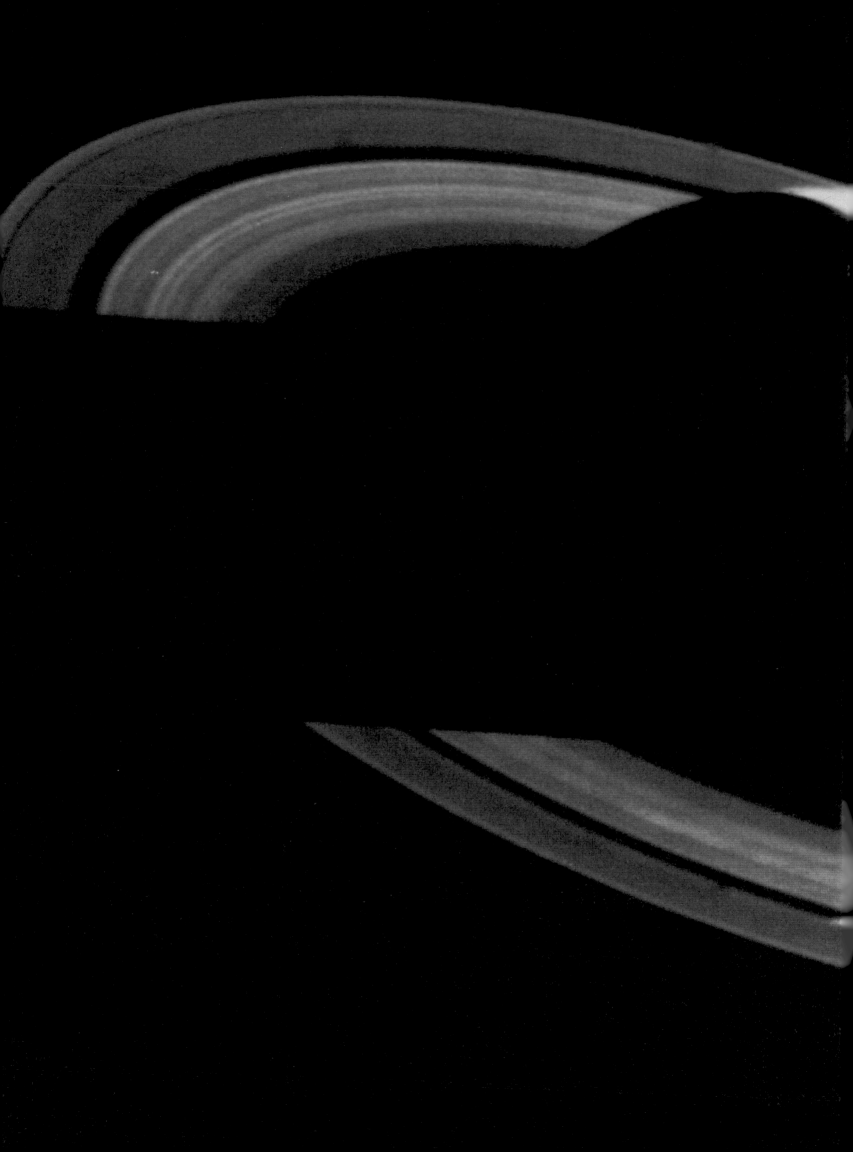

Voyager 1 looked back at Saturn on 16 November, 1980, four days after the spacecraft flew past the planet. A few of the spokelike rings discovered by Voyager appear in the rings as bright patches. From Saturn, Voyager 1 is on a trajectory taking the spacecraft out of the ecliptic plane, away from the Sun and eventually (by 1990) out of the solar system.

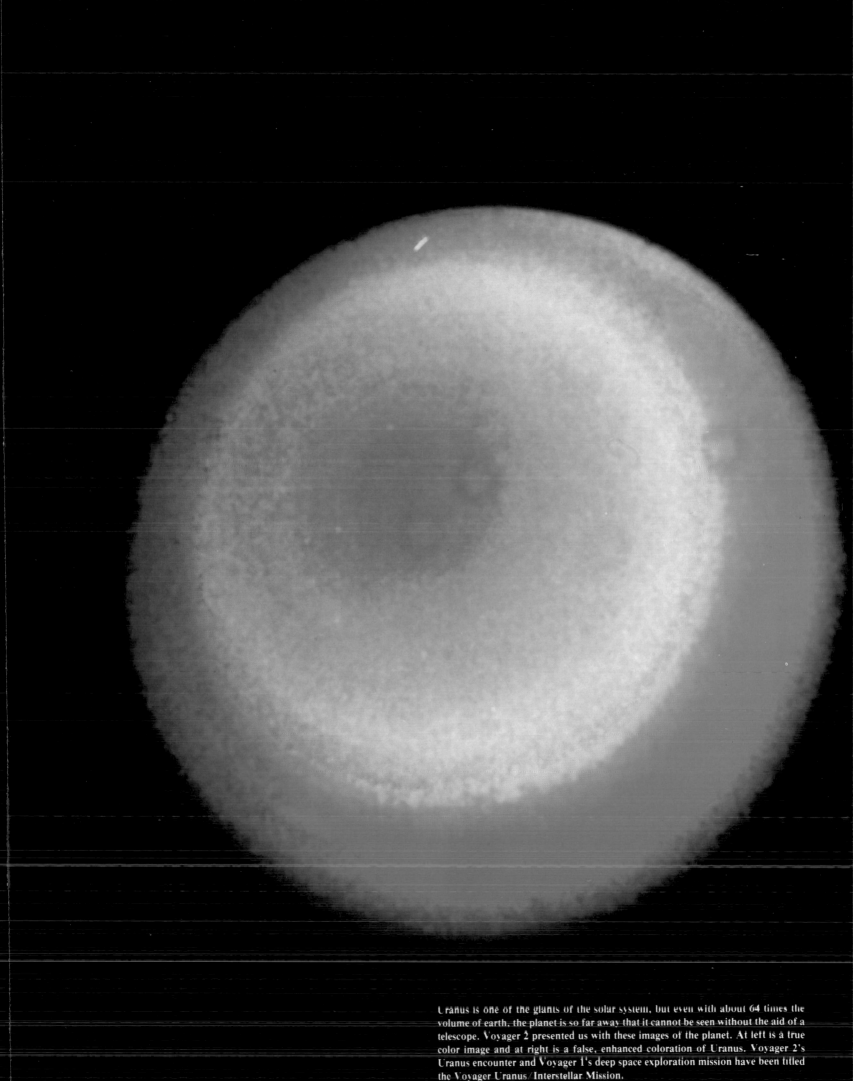

Uranus is one of the giants of the solar system, but even with about 64 times the volume of earth, the planet is so far away that it cannot be seen without the aid of a telescope. Voyager 2 presented us with these images of the planet. At left is a true color image and at right is a false, enhanced coloration of Uranus. Voyager 2's Uranus encounter and Voyager 1's deep space exploration mission have been titled the Voyager Uranus/Interstellar Mission.

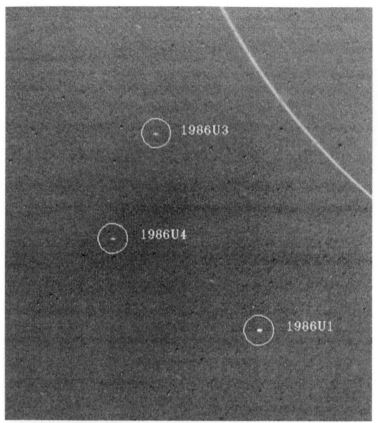

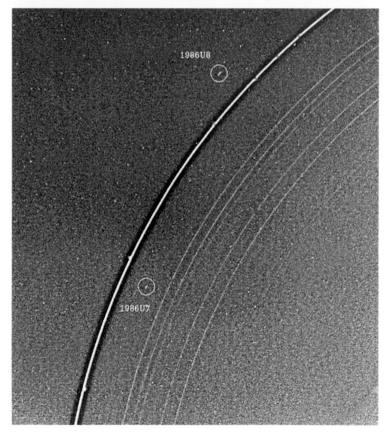

(*Above, top*) Three of the moons discovered by Voyager 2: 1986U1, 1986U3, 1986U4, with an epsilon ring. Ten new moons were discovered by Voyager 2 positioned between the rings and the larger, known moons beginning with Miranda. One of the moons was found to be nestled in the rings. (*Above*) Two 'shepherd' moons, 1986U7 and 1986U8, with epsilon ring. The existence of the shepherd moons confirmed the theory of Goldreich and Tremaine that the rings of Uranus must be supported by shepherd moons in order to keep from drifting apart.

(*Right*) All of the known rings of Uranus including newly discovered 10th ring designated 1986U1R (barely visible below outermost, epsilon ring). Uranus' rings are held together by the shepherd moons which orbit directly inside and outside the rings and exchange gravitational energy with the particles from the rings.

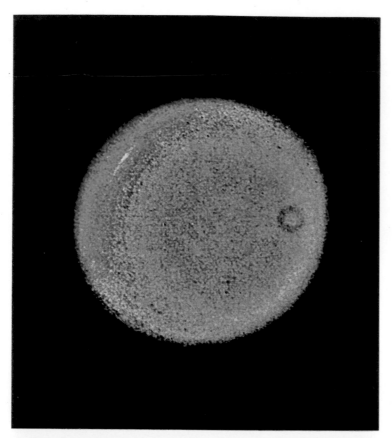

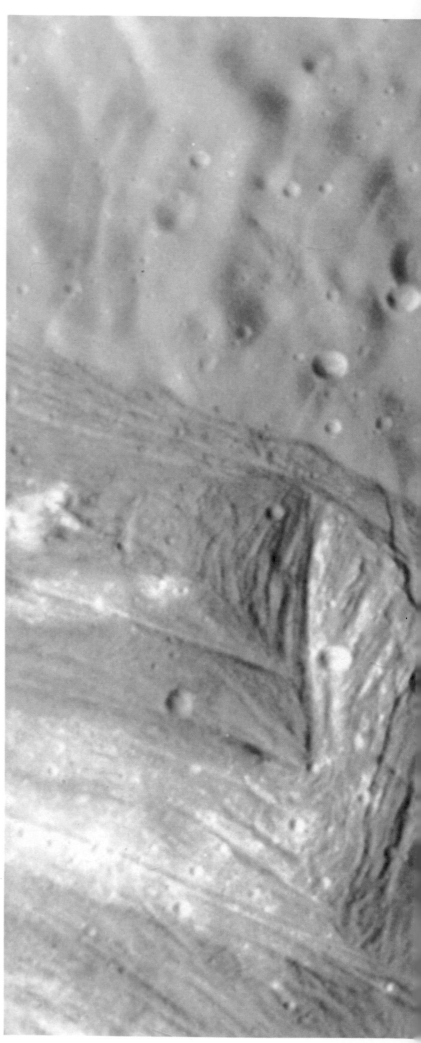

(*Above, top*) A false-color composite of Uranus shows a discrete cloud in its upper atmosphere. (*Above*) Titania, 272,000 miles from Uranus, exhibits the effects of the severe geological activity that formed the Uranian moons. This high-resolution photograph taken 24 January 1986 displays prominent fault valleys nearly 1000 miles long. (*Right*) Miranda, 80,000 miles from Uranus, revealed the most interesting landscapes of the Voyager's mission to Uranus. This photograph depicts the 'chevron' figure as seen by Voyager on its approach to Miranda from a range of 26,000 miles. Intense and violent collisions of comets, asteroids, and perhaps other moons with Miranda appear to have been the source of these geological features. Miranda may have been impacted with such force that it literally broke up under the strain and was left to reamalgamate its matter repeatedly, using its own gravitational forces.

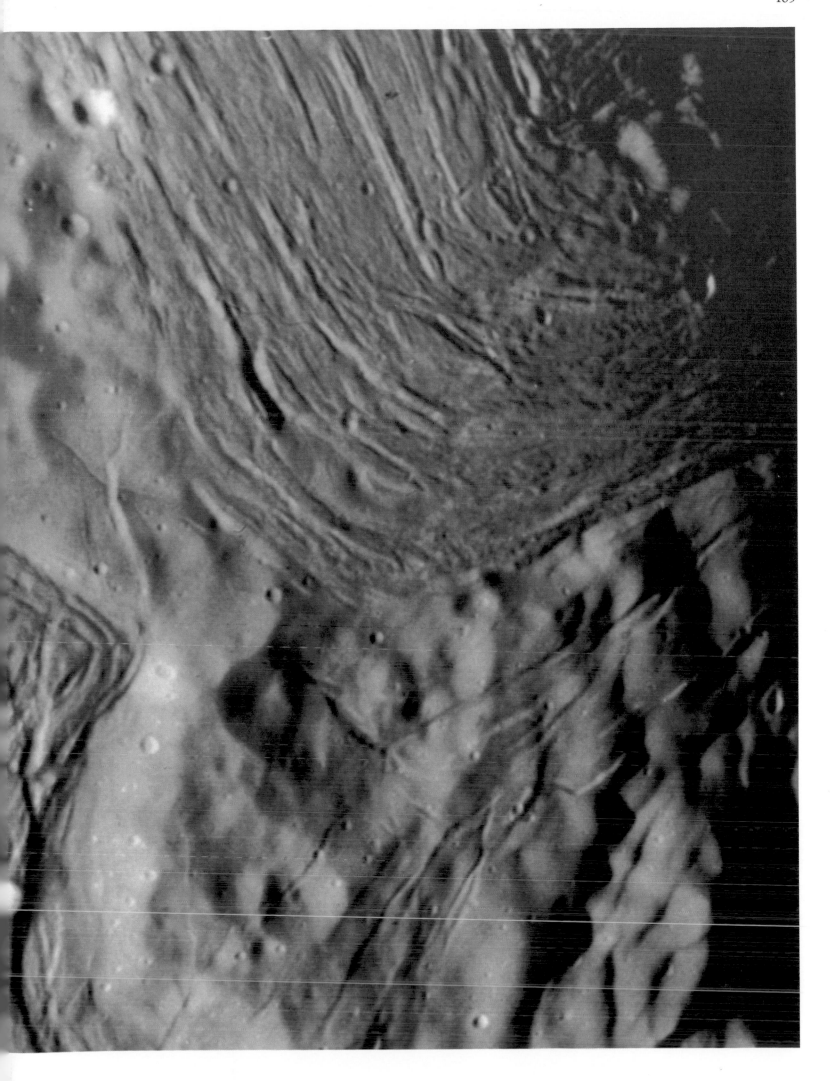

(*Above, top*) Oberon, 364,000 miles from Uranus, is the furthermost of Uranus' moons. This image of Oberon shows cratering and a large peak on the moon's lower limb. (*Above*) This high-resolution photograph of Ariel, 119,000 miles from Uranus, was the most detailed view received from Voyager 2, and shows numerous faults and valleys. The moons nearer to Uranus exhibit more turbulence than those that are a greater distance away. (*Right*) Miranda displays rugged, high-elevation terrain and lower, grooved terrain and a large crater 15 miles across — indications of its complex geologic history. This picture depicts the so-called 'pancake' that may be an indication of the reamalgamation Miranda has undergone. Denser material, which should have settled to the center of the body long ago, has been brought to the surface by the strong collisions with other bodies and has caused this geological feature to appear.

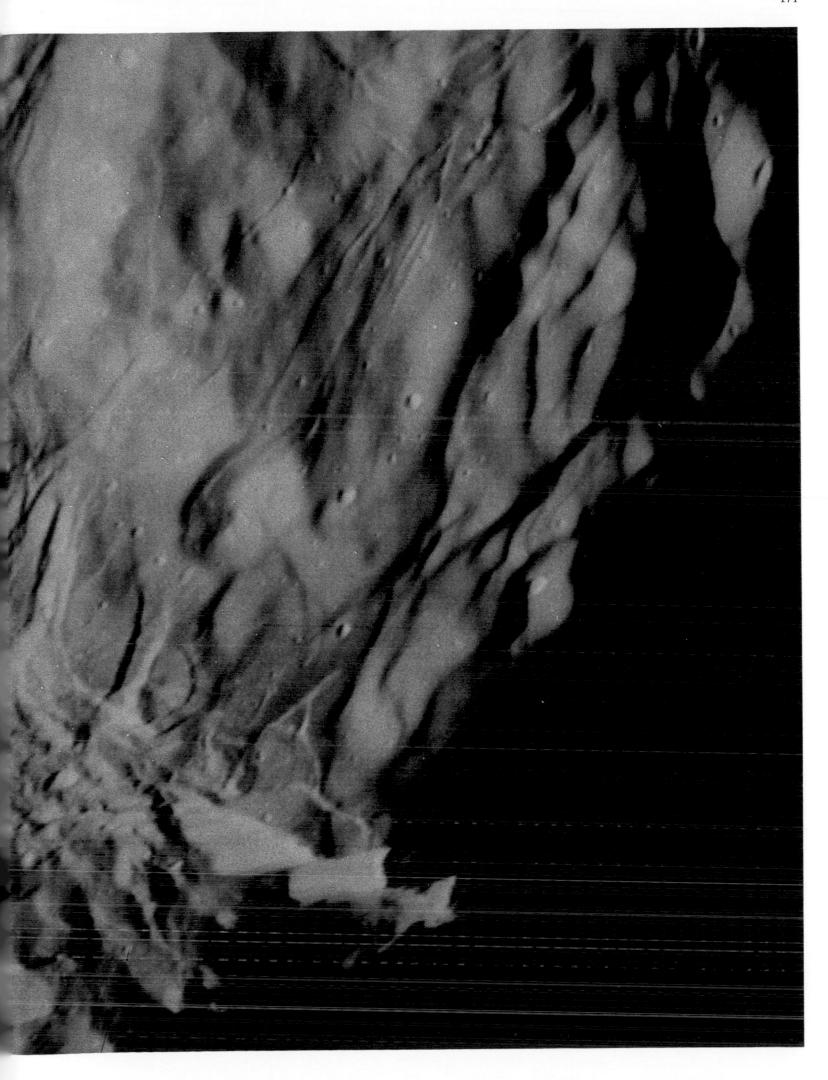

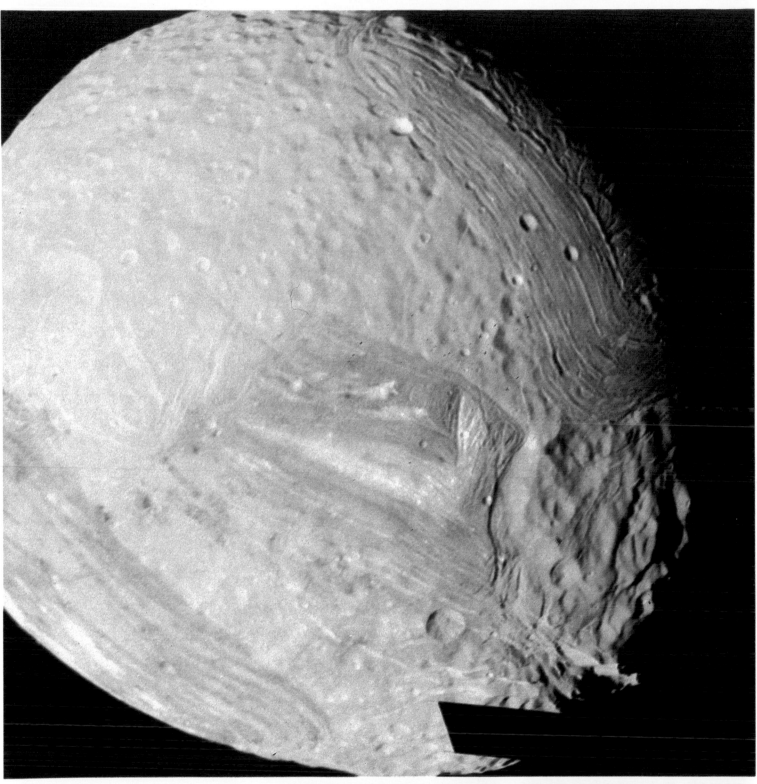

(*Left*) This montage simulates the view over horizon of Miranda toward Uranus and its rings 65,000 miles away. (*Above*) A computer mosaic of Miranda images shows varied geologic regions at high-resolution. In this photograph, the difference of the density of matter at the surface can be clearly detected, and Miranda's geological formations are shown as they geographically relate to each other. The 'chevron' is near the middle of the planet, the 'pancake' is below and to the right of the 'chevron', and the 'race track' stretches across the left side of Miranda.

(*Right*) A close-up of the 'race track', another extremely unusual geological feature on Miranda. (*Far right*) A Voyager 2 image of Miranda taken shortly before its closest approach to the planet. At the point of closest approach, on 24 January 1986, the spacecraft flew 50,600 miles above Uranus' cloud tops.

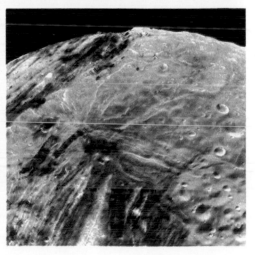

(*Above*) This backlit view shows continuous distribution of fine particles throughout the ring system. The amount of dust particles found by Voyager 2 was considerably less than expected, presumably due to the unusual nature of Uranus' magnetic field and its rotation axis, which is almost opposite that of Earth. (*Left*) A farewell shot of crescent Uranus as Voyager plunges into the dark black void on the three-year odyssey that will take it to Neptune.

The Hubble Space Telescope

Placing a telescope in space has been a dream of astronomers for decades, dating back to before space flight was a fact. There, far above the distorting effects of the earth's atmosphere, astronomers would have an unimpaired view of the entire universe. NASA is meeting this dream of astronomers through the Edwin P Hubble Space Telescope, a national observatory orbiting 320 nautical miles above the earth.

The Hubble Telescope is a 94.5-inch Ritchey-Chrétien telescope. It is named in honor of American astronomer Edwin P Hubble, who made vital contributions to the understanding of galaxies and the universe through his work earlier in this century. While the mirror size and optical quality make the Hubble Telescope one of the largest and most precise astronomical instruments ever produced. Its major advantage is that it will be outside the earth's atmosphere. The telescope's high vantage point will allow it to see farther and with greater clarity than any astronomical instrument ever built. Besides 'seeing' better, it will detect more portions of the electromagnetic spectrum, such as ultraviolet light, which is absorbed by the atmosphere before reaching the ground.

The heart of the Space Telescope is the Optical Telescope Assembly (OTA). The major segments of the OTA are the 94.5-inch primary mirror, a secondary mirror of 12 inches and the OTA's support structure. The precision of the primary and secondary mirror is a major ingredient in the superb capability of the Space

Telescope. If the mirror were scaled up to the size of the earth, none of the great mountain ranges would tower more than 5 inches above the lowest point. Light entering the Space Telescope is reflected off the primary mirror to the secondary mirror, 16 feet away. The secondary mirror sends the light through a hole in the center of the large mirror, back to the scientific instruments.

Providing all the essential systems to keep the Hubble Telescope operating in the hostile environment of space is the function of the Support Systems Module (SSM). The SSM also directs communi-cations, commands, power and fine pointing control for the tele-scope. Collectively, the SSM consists of the light shield on the front end of the telescope; the equipment section, with the main spacecraft electronics equipment; and the aft shroud, which contains the scientific instruments.

The Fine Guidance Sensors (FGS) feed roll, pitch and yaw infor-mation to the telescope's attitude-control system. The pointing capability of the Hubble Telescope provided by the FGS is so precise it is often called a sixth scientific instrument. To point the telescope, the FGS must identify the position of specified stars. This pointing data can be used to calibrate space-distance relationships throughout the universe. The FGS will allow the telescope to point with a stability of 0.007 arc seconds, or roughly the equivalent of

focusing on a dime in Los Angeles from a vantage point in San Francisco.

The Hubble Space Telescope carries five scientific instruments, which are replaceable and serviceable in orbit. Four of the instruments, about the size of telephone booths, are located in the aft shroud, behind the primary mirror. They receive light directly from the secondary mirror. The fifth instrument, the Wide Field/Planetary Camera, is located on the circumference of the telescope and uses a pick-off mirror system.

The Faint Object Camera (FOC) does exactly what its name implies; it observes faint objects. It does this by taking very low light levels and electronically intensifying the images. Objects as faint as 28th or 29th magnitude (the higher the magnitude, the fainter the object) should be observed by the FOC. By comparison, Earth-based telescopes can see to about 24th magnitude. Likely targets for the instrument are the search for extrasolar planets, variable

brightness stars, and in its spectrographic mode, the center of galaxies suspected of concealing black holes.

The Faint Object Spectrograph (FOS) will measure the chemical composition of very faint objects. Visible light contains information used to determine the chemical elements that make up the light source. Special gratings and filters allow the FOS to make spectral exposures that not only reveal information about the makeup of a light source but also about its temperature, motion and physical characteristics. This instrument will study the spectra of objects in the ultraviolet and visible wavelengths. Particular targets of interest are quasars, comets and galaxies.

While performing in much the same way as the FOS, the High Resolution Spectrograph (HRS) will observe only the ultraviolet portion of the spectrum. Ultraviolet light is filtered by the earth's atmosphere, and the only spectrographic measurements taken in this light have been from previous space observatories. The HRS

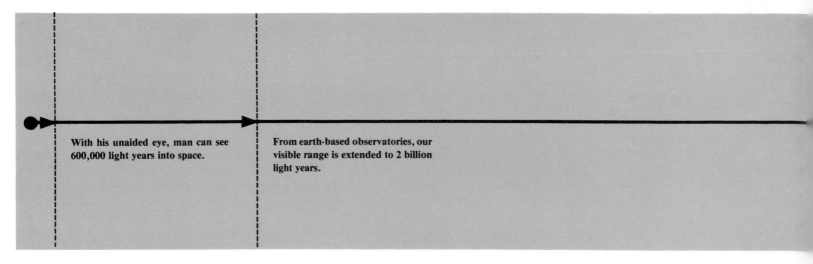

With his unaided eye, man can see 600,000 light years into space.

From earth-based observatories, our visible range is extended to 2 billion light years.

will be used to investigate the physical makeup of exploding galaxies, interstellar gas clouds and matter escaping from stars.

With no moving parts, the High-Speed Photometer (HSP) is the simplest of the five Space Telescope instruments. It will allow astronomers to take very exact measurements of the intensity of light coming from stellar objects. In addition, it will provide very precise measurements, down to the microsecond level, of time variations in the light. The amount of light received from an object is an important factor in determining its distance, making the HSP useful in refining the scale of the Milky Way galaxy and other nearby galaxies.

Actually two separate cameras in one housing, the Wide Field/Planetary Camera (WF/PC) should return some of the most spectacular visual images from the Hubble Space Telescope. In the Wide Field mode the instrument will view large areas of space and provide exquisite views of galaxies and star fields. In the Planetary

(Above and left) Two possible methods for deploying the Hubble Space Telescope from the space shuttle. After the shuttle launches the Hubble telescope, it can also serve as a base for making repairs by replacing modular components. The Shuttle may also bring the telescope back to Earth for maintenance or overhaul. (*Overleaf*) The Hubble Space Telescope in the Lockheed factory at Sunnyvale, California.

mode, it will provide glimpses of the planets comparable to those obtained on close flyby missions.

The Hubble Space Telescope will be placed in orbit by the Space Shuttle in the autumn of 1987. Its orbit will be 320 nautical miles high and inclined to the equator 28.5 degrees. After being deployed from the Shuttle cargo bay, the solar arrays and high-gain antennas will be deployed. The shuttle will station-keep nearby while final checks are completed. After a period of calibration and verification, astronomers at the Space Telescope Science Institute will begin their observations of the sky. Provisions have been made for on orbit repair and maintenance of the Space Telescope. Astronauts

With the Hubble Space Telescope, the astronomers lucky enough to use it will be able to see for 14 billion light years. If the universe is younger than 14 billion years, as some have theorized, will those astronomers then be able to see its creation?

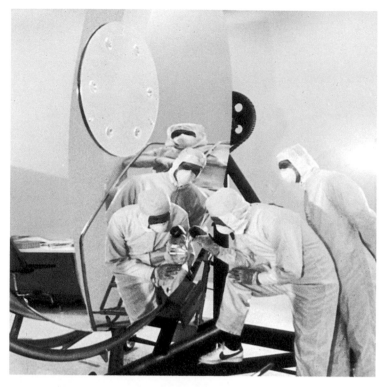

(*Above*) Engineers inspecting the 94.5-inch primary mirror of the telescope. Light enters the front end, is projected from the primary mirror to the secondary mirror, then is diverted to a focal plane. (*Right*) Lowering the telescope into the canister that will carry it into space.

can remove old instruments and replace them with more advanced or different types of devices. Repairs to many of the systems or instruments can also be accomplished on orbit. The telescope can be returned to Earth for major refurbishment and alteration if necessary. The operating life of the Hubble Space Telescope may be 15 years or more with refurbishment and modernizing using the Space Shuttle.

The Office of Space Science and Application at NASA Headquarters is responsible for overall program management, financial and scheduling provisions, and the science policy development and direction. Marshall Space Flight Center in Huntsville, Alabama is responsible for the development and operation of the Space Telescope system as the 'lead' NASA center. In Greenbelt, Maryland, the Goddard Space Flight Center is managing the scientific instruments, mission operations and data management. It is also charged with monitoring the Space Telescope Science Institute.

The Space Telescope Science Institute is operated by AURA, the Association of Universities for Research in Astronomy. The Institute is located on the Homewood Campus of Johns Hopkins University in Baltimore, Maryland. It is the job of the Institute to determine the observational program of the Space Telescope while on orbit, ensuring that the observatory will be used to its maximum advantage. Lockheed Missiles & Space Company, Sunnyvale, California, is the systems integrator of the Space Telescope satellite. It is also responsible for the design, development and manufacture of the Support Systems Module. Perkin-Elmer Corporation of Danbury, Connecticut manufactured the Optical Telescope Assembly.

The Hubble Space Telescope

Overall length	13.1 m	43.5 ft
Overall diameter	4.3 m	14.0 ft
Weight	11,000 kg	25,200 lb
Primary mirror (diameter)	2.4 m	94.5 in
Secondary mirror (diameter)	0.3 m	12.2 in

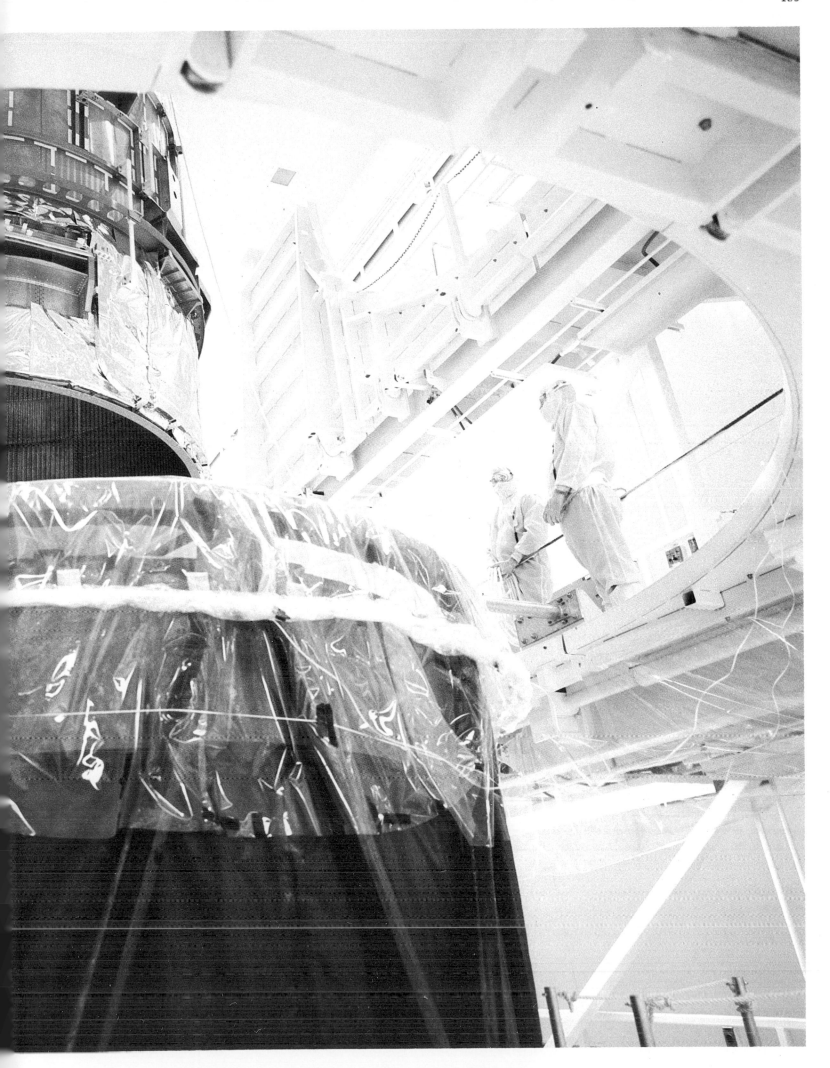

(Right) An artist's view of the Hubble Space Telescope in operation. Freed from the Earth's murky atmosphere and equipped with an unparalleled optical sytem, the telescope will provide an unprecedented opportunity for viewing the universe.

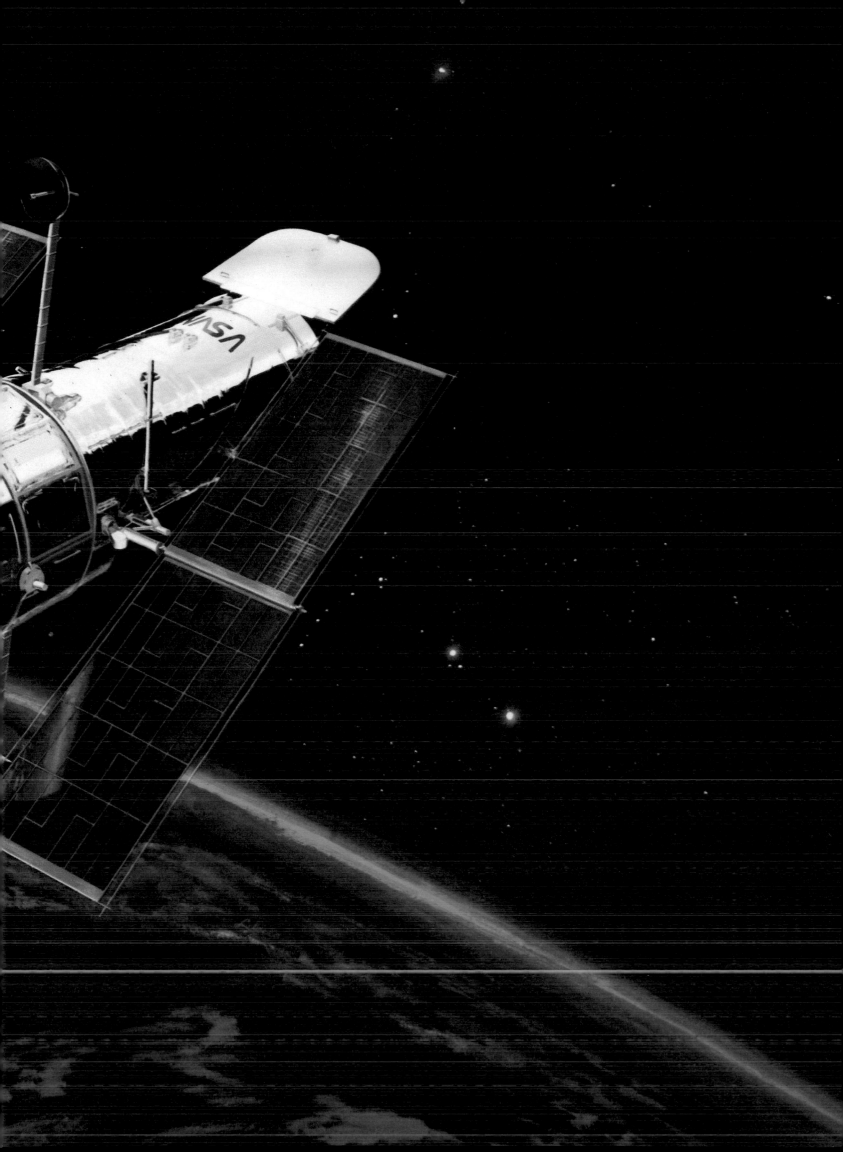

(*Left*) The Galileo Probe in an artist's depiction of its journey to Jupiter. The Probe will separate from its parent Orbiter for its descent and destruction into the atmosphere of Jupiter while the Orbiter conducts its own studies of Jupiter, its magnetosphere and its major satellites. The Galileo Probe's launch has been postponed until 1987.

(*Left*) Sunrise on Venus as seen from the Pioneer Venus Orbiter in 1978. The Pioneer spacecraft were one of the early triumphs of NASA's interplanetary exploration and a candidate for future investigations. In this view, the planet's surface is entirely blocked from view by a thick veil of remarkably uniform bright haze

INDEX

Advanced X-Ray Astronomy Facility 33
Aldrin, Buzz 22
Alouette Missions 40
Anders, William 20
Andromeda Galaxy *50, 51*
Apollo 20, *20-21,* 22, *22-23,* 36
Archimedes crater 20
Argyre Planitia *102*
Ariel *170*
Aristarchus 8
Armstrong, Neil 22
Association of Universities for Research in Astronomy (AURA) 182
Asteroid Belt 84, 86, 87
Astron 30, 32
Aztec Calendar Stone 7

B-ring 158, *192-193*
Bean, Alan *20-21*
'Big Joe' *111*
Borman, Frank 20
Bunsen, Robert 12

C-ring *192-193*
California Institute of Technology 33
Callisto 38, 84, 121, *130, 136-137*
Canada-France-Hawaii Telescope *24-25,* 25
Canopus 69, 71, 72, 73
Cassini division of Saturn 158
Cen X-3 32
Challenger 33, 57
Chryse Planitia *116-117*
Circinus 47
Cirrus clouds *62-63*
Comet Bennett 60
Comet Giacobini-Zinner 43
Comet Tago-Sato-Kosaka 43, 60
Conrad, Charles *20-21*
Copernicus, Nicolaus 8, 30, 43
Copernicus see Orbiting Astronomical Observatories
COS-B Satellite 33
Crab Nebula *47*
Crab pulsar 33
Cygnus constellation 46, 47
Cygnus X-1 32, 43, 46
Cygnus X-3 32

Deimos 22, 77, 104
Dione *148-149, 153, 154-155, 161*
Dollond, John 8

Eagle Nebula *28-29*
Earth 6, 8, 20, 34, 36, 38, 40, 43, 46, 48, 55, 57, 60, *61,* 72, 77, 79, 81, 84, 117, 126, 139, 144
Einstein Observatory *see* High Energy Astronomy Observatories
Enceladus 152, *154*
Equatorial telescope *12*
Europa 22, 38, 84, 121, *128-129, 130, 131, 135*
European Space Agency (ESA) 30, 32, 33, 40, 43
European Space Research Organization *see* European Space Agency
Exosat Observatory 32
Explorer spacecraft 30, 32, 33, 38, *39,* 40, 41, *42,* 43, 59, 60

F-ring 22, *158-159*

Galileo, Galilei *8,* 12, 20, 33, 38
Galileo Probe 33, *186-187*
Ganymede 38, 84, 121, 126, *130, 132-133, 140, 141,* 154
Gegenschein 67
Geminga 33
Gemini 33
Goddard Space Flight Center 182
Great Nebula *14-15*
Great Red Spot 22, 84, 86, *127, 134,* 135, *138, 139*
Griffith Observatory *12-13*

Hakucho Observatory 32
Hale, George Ellery 17
Hall crater *104*
Halley's Comet *32,* 33, 57
Hamilton, Mount *18-19*
Helios spacecraft 60
Hellespontus 73
Hercules 47
Herschel, William 9
High Energy Astronomy Observatories 32, 46, 47, 48
Hipparchus 7, 8
Hooker reflector 17
Hubble, Edwin P 17, 33, 176
Hubble Space Telescope 25, 33, 176-185, *177-185*
Hughes Aircraft 57, 82
Hyperion 152

Iapetus 86, 152
Infrared Astronomy Satellite (IRAS) 27, 30, 33, *35,* 48, *48, 49,* 51, 53, 55, *57, 60-61,* 61, *62-63*
International Magnetospheric Study 43
International Sun-Earth Explorer C Satellite (ISEE) *see* Explorer spacecraft
International Ultraviolet Explorer (IUE) *see* Explorer spacecraft
Io 22, 38, 81, 84, *86,* 121, *130, 134, 135, 144, 145, 146-147*

Jansky, Karl 17
Jet Propulsion Laboratory 55
Johns Hopkins University 182
Jupiter 8, 22, 36, 38, 46, 67, 84, 86, 121, 123, 125, 126, *126, 130, 134, 138, 139, 142-143,* 144, *186-187*

Kepler, Johannes 8
Kepler Ridge *104*
Kirchhoff, Gustav 12
Kitt Peak National Observatory *10-11,* 12, *26-27*
Kuiper Airborne Observatory (KAO) 27, 60

Lagoon Nebula *10-11*
Lick Observatory *18-19*
Lockheed Missiles and Space Company 182
Lovell, James 20
Lowell, Percival 101
Luna Spacecraft 20
Lunar Orbiter *36-37*
Lunar Rover *22-23*

Magellanic Clouds 48, *60-61*
Mariner spacecraft 22, *73,* 97
Mariner 1 67
Mariner 2 22, 67
Mariner 3 69
Mariner 4 22, 67, *68-69,* 71, 72, 73
Mariner 5 *70-71,* 71
Mariner 6 72, 74, 77
Mariner 7 72, 77
Mariner 8 74, 75
Mariner 9 22, 74, *74-75,* 75, 77, 97
Mariner 10 22, 76, 77, *78-79,* 79
Mars 17, 22, 36, 46, 64, 67, 69, 71, 72, 73, *74,* 74, 75, 77, 81, 87, 88, *90-91,* 97, *101,* 102, 104, *106-107, 108-109, 111,* 113, *116-117,* 117, *118-119*

Marshall Space Flight Center 182
Mauna Kea *24-25,* 25, 27, 33
Mayall, Nicholas U Telescope *27*
Mercury 22, 36, 38, *76-77,* 77, 79, 126, 141
Meridianii Sinus 73
Milky Way 17, 30, 32, 38, 40, 48, *52, 53,* 60, *61*
Mimas 152, *161*
Miranda *166, 168-169, 170-171, 172, 173*
Moon 8, *16,* 17, 20, *20-21,* 22, *22-23,* 36, 41, *42,* 73, 79, 141
Mutch, Dr Thomas A 88
Mutch Memorial Station *see* Viking 1 Lander

NASA (National Aeronautics and Space Administration) 33, 34-175, 176, 182
NASA Eastern Test Range 59
NASA Infrared Telescope Facility *24-25,* 25
NASA Western Test Range 55
Neptune 25, 38, 61, 84, 86, 125, 144
Nelson, George 57, *59*
Netherlands Astronomical Satellite (NAS) 48
Newton, Isaac 8, 9, 12
Nix Olympica 73
Nova Serpentis 43

Oberon *170*
Olympus Mons *104*
Orbiting Astronomical Observatories (OAO) 30, 43, *44-45,* 45, 46, 60
Orbiting Geophysical Observatories (OGO) *57,* 60
Orbiting Solar Observatories *56,* 57
Orion constellation *14-15,* 30
Orion nebula *54*

Palomar, Mount 17, 33, 50, 52, 126
Perkin-Elmer Corporation 82
Phobos 22, 77, *104*
Phoebe 38, 121
Pioneer Spacecraft, early missions 22, 64, 65, 67
Pioneer 10/11 22, 64, *65,* 67, 84, *84, 84-85,* 86, 126
Pioneer Venus Project 22, *80,* 81, 82, *82,* 83, *83,* 189
Pluto 22, 38, 60, 61, 64, 84, 86, 126
Princeton Experiment Package (PEP) 43

Project SETI (Search for Extra-Terrestrial Intelligence) 61
Ptolemy 7, 8

Radio Astronomy Satellites *see* Explorer spacecraft
Ranger spacecraft 20, *66,* 67
Reber, Grote 17
Rhea 86, *148-149,* 152, *161*

Sagittarius *9*
Saturn 9, 22, 36, 38, 46, 86, 121, 123, 125, 126, 144, *148-149, 150-151, 153, 158-159, 160, 161, 162-163, 192-193*
Schmidt, Harrison 22, *22-23*
Schweikert, Rusty 20
Scorpius 32, 47
Scorpius X-1 30
Serpens Constellation *28-29*
Shuttle Infrared Telescope Facility 33
Small Astronomical Satellites (SAS) *see* Explorer spacecraft
Solar Maximum Mission (SMM) 57, *58,* 59, *59*
'Sounds of Earth' 121, 126
Space Shuttle 25, 179, 182
Space Telescope *see* Hubble Space Telescope
Space Telescope Science Institute 179, 182
Spectroscopy *9,* 12
Sputnik 17
Stargazer see Orbiting Astronomical Observatories
Stickney crater *104*
Sun 8, 12, 17, 30, 32, 34, 35, 38, 40, 43, 46, 48, 57, 59, 60, 64, 65, 86
Surveyor spacecraft 20, *20-21*

Taurus Littrow *22-23*
Tenma Observatory 32
Terzan 2 47
Tethys *148-149,* 152, *152, 153, 161*
Titan 38, 86, 104, 126, *154, 156-157, 161*
Titania *168*

Uhuru see Explorer spacecraft
United Kingdom Infrared Observatory *24-25,* 25
United States Navy 38
University College of London Experiment Package 43

University of Colorado 40, 57
University of Iowa 40
Uranus 9, 22, 25, 27, 38, 60, 86, 125, 144, *164-165, 166-167, 168, 172, 174-175*

Valles Marineris *92-93, 106-107*
Van Allen Radiation Region 34, 38, 40, 41, 60, 67
Vanguard spacecraft 38
van Hoften, James 57, *59*
Vega 30
Vela 33
Venera 7, 22
Venus 8, 17, 22, 36, 67, 71, 77, 79, 81, 82, *82,* 83, 97, *188-189*
Viking Spacecraft 22, *98-99,* 101
Viking 1 Lander 22, 88, *94-95,* 97, *99,* 101, 109, *111, 113,* 117
Viking 1 Orbiter 88, 97, *97,* 101, 107
Viking 2 Lander 22, 88, 101, 111
Viking 2 Orbiter 88, *88-89,* 101
von Fraunhofer, Joseph 12
Voyager spacecraft 22, 25, 86, *120,* 121, *122-123, 124-125,* 139, 143, *144,* 152, 165
Voyager, encounters with Jovian System 22, 121, 125, 126, *126,* 129, 136, 141
Voyager, encounters with Saturnian System 22, 38, 121, 125, 126, *126,* 149, 152, 158, 163, 191
Voyager, encounter with Uranian System 22, 25, 125, 126, 144, 152, 173
Voyager Uranus/Interstellar Mission *see* Voyager, encounter with Uranian system

Wilson, Mount 17

Yerkes Observatory 9, 17

1986U1 (Uranian moon) *166*
1986U1R (Uranian ring) *166-167*
1986U3 (Uranian moon) *166*
1986U4 (Uranian moon) *166*
1986U7 (Uranian moon) *166*
1986U8 (Uranian moon) *166*

(Overleaf) **Saturn's C-ring and B-ring with many ringlets as seen by Voyager 2 during its voyage to Jupiter, Saturn, Uranus and points beyond. Voyager 2 flew by Saturn in August of 1981 to record hundreds of important images.**